Critical Analysis of World Uranium Resources

By Susan Hall and Margaret Coleman

Prepared in cooperation with the U.S. Energy Information Administration

Scientific Investigations Report 2012–5239

U.S. Department of the Interior
U.S. Geological Survey

U.S. Department of the Interior
KEN SALAZAR, Secretary

U.S. Geological Survey
Marcia K. McNutt, Director

U.S. Geological Survey, Reston, Virginia: 2013

For more information on the USGS—the Federal source for science about the Earth, its natural and living resources, natural hazards, and the environment, visit *http://www.usgs.gov* or call 1–888–ASK–USGS.

For an overview of USGS information products, including maps, imagery, and publications, visit *http://www.usgs.gov/pubprod*

To order this and other USGS information products, visit *http://store.usgs.gov*

Suggested citation:
Hall, Susan, and Coleman, Margaret, 2013, Critical analysis of world uranium resources: U.S. Scientific Investigations Report 2012–5239, 56 p.

Contents

Figures

Appendix Figures

Tables

Appendix Table

Conversion Factors, Definitions, and Abbreviations

Metric units are used throughout this report because international reporting uses these terms. Uranium resources are reported as metric tonnes [of] contained uranium metal (tU). In discussions of rates of use, the abbreviation tU/yr denotes metric tonnes of uranium per year. In the U.S. domestic mining industry, the more commonly used unit is pounds (lbs.) uranium oxide (U_3O_8). For convenience, 1 metric tonne is usually written 1 tonne. International usage reports uranium concentration in ore as percent U (uranium metal), not percent U_3O_8 (uranium oxide).

Useful conversion factors are:

1 short ton U_3O_8 = 0.769 tU

1 percent U_3O_8 = 0.848 percent U

Cost categories are reported in U.S. Dollars (USD). Converting U.S. pounds into kilograms, the price in U.S. dollars per pound (= 2.6 kilograms (kg)) of uranium oxide is written as equivalent to the price in U.S. dollars per kg of uranium metal:

1 USD (U.S. Dollar)/lb. U_3O_8 = USD 2.6/kgU

Uranium resources are reported in categories that are based on the degree of geological assurance for, and the economic feasibility of, extraction of the ore. This report states resources in terms of being economically feasible to extract and that have been explored sufficiently such that the limits and geological extent of the properties of the deposit are well constrained, typically by drill data. Throughout this report, the term Reasonably Assured Resources (RAR) is used to describe this class of resource. Other classification systems assign different terms to such resources. This report's usage corresponds roughly to the U.S. Department of Energy (DOE) classification "reserves recoverable at less than USD 50 per pound uranium oxide" (<USD 50/lb. U_3O_8), equivalent to USD 130 per kilogram uranium metal (USD 130/kgU); it corresponds to and indicated reserves in the U.S. Geological Survey (USGS) classification system; to economic demonstrated resources in the Australian national scheme; to proved and probable reserves as used by Australasia's Joint Ore Reserves Committee (JORC); and to measured and indicated resources of the Canadian Institute of Mining, Metallurgy and Petroleum (CIM) (fig. 1).

Decreasing degree of geological assurance →

Decreasing degree of economic feasibility ↓

		Identified Resources			Undiscovered Resources
		Demonstrated		**Inferred**	
		Measured	Indicated		
Economic		Proved Resources RAR @ <$80/kg U Reserves, **Reserves @ <$30/lb U_3O_8** Economic Demonstrated Resources Commercial Projects	Probable Reserves RAR @ <$80/kg U Reserves, **Reserves @ <$30/lb U_3O_8** Economic Demonstrated Resources Commercial Projects	Inferred Resources **Estimated Additional Resources** (Inferred Resources Recoverable @ <$80/kg U) Inferred Reserves Economic Inferred Resources	
Economic		Measured Resources RAR @ <$130/kg U Reserves, **Reserves @ $30–$50/lb U_3O_8** Economic Demonstrated Resources Commercial Projects	Indicated Resources RAR @ <$130/kg U Reserves, **Reserves @ $30–$50/lb U_3O_8** Economic Demonstrated Resources Commercial Projects		
Subeconomic Paramarginal / Marginally Economic		Marginal Reserves RAR @ $130 to $260/kg U **Reserves @ $50–$100/lb U_3O_8** Paramarginal Demonstrated Resources Non-commercial Projects		Inferred Resources **(Inferred Resources Recoverable @ $80 to $130/kg U)** Inferred Marginal Reserves Paramarginal Inferred Resources	Hypothetical ⟶ Speculative Prognosticated ⟶ Speculative **Estimated Additional Resouces** ⟶ **Speculative Resources**
Subeconomic Submarginal / Subeconomic		Demonstrated Subeconomic Resources RAR @ $130 to $260/kg U **Reserves @ $50–$100/lb U_3O_8** Submarginal Demonstrated Resources Non-commercial Projects		Inferred Resources **(Inferred Resources Recoverable @ $130 to $260/kg U)** Inferred Subeconomic Resources Submarginal Inferred Resources	Potential Projects

Other Occurences	Includes nonconventional and low-grade materials

All schemes excluding IAEA

IAEA Scheme - RAR - Reasonably Assured Resources (losses resulting from mining/milling deducted, and cost categories from 2009 Red Book)

JORC (Australian Joint Ore Reserves Committee) Scheme

JORC (Australian Joint Ore Reserves Committee) & CIM (Canadian Institute of Mining) Schema

U8&8 Scheme (UQQ0 Circular 011,1990)

DOE - Note: DOE does not include inferred resources when reporting reserves, but includes them as part of Estimated Additional Resources (EAR)

Australian National Scheme (losses resulting from mining and milling deducted; Australia reports a category "Accessible Economic Demonstrated Resources" which subtracts deposits that cannot be currently mined—national parks, etc.)

Canada - Natural Resources Canada

UNFC (United Nations Framework Classification for Fossil Energy and Mineral Reserves and Resources) adds a third dimension to resource classifications, "E" - Economic and social viability (reasonable prospects for eventual economic extraction) in addition to "G" - geologic knowledge, and "F" project status and feasibility.
Available at: http://www.unece.org/energy/se/pdfs/UNFC/oct09/ECE.ENERGY.GE.3.2009.L5_draft_e.pdf
The system does not correlate completely with other schema, but its general correlation with other systems is shown above.
For Inferred Resources, UNFC would require deposits to be classified by numeric coding from 1 (most viable) to 4 (least viable)

$: U.S. Dollars

Figure 1. Schemes for classifying geological assurance of uranium resources, as used by six international agencies concerned with the mining of ores.

Less assured resources—those in the Inferred category, those that are subeconomic, or those that are hypothetical or as yet undiscovered—are included in this analysis only in sections of the text that project future uranium supply beyond the approximately 20-year timelines that are customary for development. Inferred resources are reported in the cost categories used by the Organisation for Economic Co-operation and Development Nuclear Energy Agency (NEA). The present report analyzes these categories when it separately examines the uranium supply in 24 countries and their scenarios for future development of supply (appendix 1). The text clearly identifies these resources as being more speculative wherever it discusses them.

Following International Atomic Energy Agency (IAEA) and Organisation for Economic Co-operation and Development, Nuclear Energy Agency (NEA) protocols, this report applied recovery factors to *in-situ* resources in order to determine the amount of uranium available after mining and processing, depending on type of mining (NEA–IAEA, 2010). Recovery can be determined with certainty only after mining is completed, because it depends on the metallurgy of the deposit and on the mining and processing methods, all of which can vary widely from deposit to deposit and during the entire course of mining. Estimated recovery adds a measure of uncertainty to analysis of the adequacy of uranium resources to satisfy demand. To minimize this bias, USGS and EIA followed standard recovery factors used by NEA and IAEA.

EIA   U.S. Department of Energy Energy Information Administration

USGS   U.S. Department of the Interior Geological Survey

NEA   Organisation for Economic Co-operation and Development - Nuclear Energy Agency

IAEA   International Atomic Energy Agency

Critical Analysis of World Uranium Resources

By Susan Hall[1] and Margaret Coleman[2]

Abstract

The U.S. Department of Energy, Energy Information Administration (EIA) joined with the U.S. Department of the Interior, U.S. Geological Survey (USGS) to analyze the world uranium supply and demand balance. To evaluate short-term primary supply (0–15 years), the analysis focused on Reasonably Assured Resources (RAR), which are resources projected with a high degree of geologic assurance and considered to be economically feasible to mine. Such resources include uranium resources from mines currently in production as well as resources that are in the stages of feasibility or of being permitted. Sources of secondary supply for uranium, such as stockpiles and reprocessed fuel, were also examined. To evaluate long-term primary supply, estimates of uranium from unconventional and from undiscovered resources were analyzed.

At 2010 rates of consumption, uranium resources identified in operating or developing mines would fuel the world nuclear fleet for about 30 years. However, projections currently predict an increase in uranium requirements tied to expansion of nuclear energy worldwide. Under a low-demand scenario, requirements through the period ending in 2035 are about 2.1 million tU. In the low demand case, uranium identified in existing and developing mines is adequate to supply requirements. However, whether or not these identified resources will be developed rapidly enough to provide an uninterrupted fuel supply to expanded nuclear facilities could not be determined. On the basis of a scenario of high demand through 2035, 2.6 million tU is required and identified resources in operating or developing mines is inadequate. Beyond 2035, when requirements could exceed resources in these developing properties, other sources will need to be developed from less well-assured resources, deposits not yet at the prefeasibility stage, resources that are currently subeconomic, secondary sources, undiscovered conventional resources, and unconventional uranium supplies.

This report's analysis of 141 mines that are operating or are being actively developed identifies 2.7 million tU of *in-situ* uranium resources worldwide, approximately 2.1 million tU recoverable after mining and milling losses were deducted. Sixty-four operating mines report a total of 1.4 million tU of *in-situ* RAR (about 1 million tU recoverable). Seventy-seven developing mines/production centers report 1.3 million tU *in-situ* Reasonably Assured Resources (RAR) (about

1.1 million tU recoverable), which have a reasonable chance of producing uranium within 5 years. Most of the production is projected to come from conventional underground or open pit mines as opposed to *in-situ* leach mines.

Production capacity in operating mines is about 76,000 tU/yr, and in developing mines is estimated at greater than 52,000 tU/yr. Production capacity in operating mines should be considered a maximum as mines seldom produce up to licensed capacity due to operational difficulties. In 2010, worldwide mines operated at 70 percent of licensed capacity, and production has never exceeded 89 percent of capacity. The capacity in developing mines is not always reported. In this study 35 percent of developing mines did not report a target licensed capacity, so estimates of future capacity may be too low.

The Organisation for Economic Co-operation and Development's Nuclear Energy Agency (NEA) and International Atomic Energy Agency (IAEA) estimate an additional 1.4 million tU economically recoverable resources, beyond that identified in operating or developing mines identified in this report. As well, 0.5 million tU in subeconomic resources, and 2.3 million tU in the geologically less certain inferred category are identified worldwide. These agencies estimate 2.2 million tU in secondary sources such as government and commercial stockpiles and re-enriched uranium tails. They also estimate that unconventional uranium supplies (uraniferous phosphate and black shale deposits) may contain up to 7.6 million tU. Although unconventional resources are currently subeconomic, the improvement of extraction techniques or the production of coproducts may make extraction of uranium from these types of deposits profitable. A large undiscovered resource base is reported by these agencies, however this class of resource should be considered speculative and will require intensive exploration programs to adequately define them as mineable. These resources may all contribute to uranium supply that would fuel the world nuclear fleet well beyond that calculated in this report.

Production of resources in both operating and developing uranium mines is subject to uncertainties caused by technical, legal, regulatory, and financial challenges that combined to create long timelines between deposit discovery and mine production. This analysis indicates that mine development is proceeding too slowly to fully meet requirements for an expanded nuclear power reactor fleet in the near future (to 2035), and unless adequate secondary or unconventional resources can be identified, imbalances in supply and demand may occur.

[1] U.S. Geological Survey.

[2] U.S. Energy Information Administration.

Introduction

The Blue Ribbon Commission on America's Nuclear Future was established by President Obama under provisions in the Federal Advisory Committee Act (5 U.S.C. App.2) in 2010 to review and recommend policies for managing spent nuclear fuel from the nuclear power industry. As part of the study, the Commission requested that the EIA assess primary uranium supply globally, and to then compare that supply with world requirements or demands. EIA, in collaboration with the USGS, responded to the request by researching and analyzing the global resource base for uranium. This paper summarizes the results of this analysis, and addresses the question of whether there is sufficient uranium to supply the present or an expanded U.S. nuclear power reactor fleet in the near term (~25 years), and beyond.

Geologists from the USGS and EIA examined information describing uranium production, resources, and issues related to the continuity of supply of uranium, from all countries that the NEA and the IAEA have identified as containing uranium resources. Reasonably assured resources (RAR), production capacity, and mine life for individual production centers of operating mines and for mines estimated to come online in the near future (~5 to 10 years) were critically examined. Determining long-term supply is more problematic: because projections are based on uranium-containing properties that have not been fully explored, their contained uranium is uncertain. Further uncertainty arises because technical, economic, or political challenges may prevent many such properties from coming into production, even though they are geologically defined. Despite uncertainties, this report uses the best information available in order to explore the potential extent of future supply, as well as the challenges that individual production centers may encounter. Although information about the individual deposits is from the best and most objective sources available, it is beyond the scope of this project for the EIA or USGS to independently verify, through site visits to uranium producers, the accuracy of all the information that the report used as the basis for its analysis.

This report reflects the state of the industry as of December 2011, modified by comments in the narrative that reflect important events that changed the world uranium supply while this paper was in review. It should be noted that the consequences of the recent nuclear accident at the Fukushima Daiichi nuclear plant are not yet fully understood so far as they may relate to the analysis of how the supply of uranium is connected with demand for the metal. To date, the accident has resulted in slightly lower or delayed projected future demand, as in the projection by the World Nuclear Association that nuclear capacity will rise from 393 gigawatts (GW) in 2009 to 630 GW in 2035, an estimate that is 20 GW lower than the Association projected before Fukushima. Most countries that are members of the OECD, and many non-OECD countries, continue operating existing and developing new nuclear powerplants, albeit delaying such development slightly as they review safety standards for new and for existing plants (World

Nuclear Association, 2011b). Another significant recent development is the suspension of the Olympic Dam mine expansion by BHP Billiton that changes forward supply projections for uranium (ABC News, 2012).

Uranium Supply and Demand Worldwide

In 2010, there were 442 nuclear powerplants operating worldwide that required 68,646 tonnes of uranium metal (tU), and 53,663 tU was mined from 16 countries satisfying 78 percent of world requirements (World Nuclear Association, 2011b). Primary sources—active mines that recover uranium as a primary product, a coproduct, or an important byproduct—and a number of secondary sources supply uranium to the world uranium market. Not currently contributing to world uranium supply are unconventional resources, such as uranium in phosphate rocks, in black shale, or in seawater. Although unconventional resources contain a large amount of uranium, the uranium is recoverable only as a minor byproduct.

Secondary sources include (1) stocks and inventories of natural and enriched uranium held in government and in private industry stockpiles, (2) reprocessed spent reactor fuel and recycled plutonium from military sources such as the United States/Russian program in which highly enriched uranium (HEU) is converted to low-enriched uranium (LEU) ("the HEU/LEU program"), and (3) uranium produced from depleted uranium tails. From 1945 through 1991, yearly production outpaced demand by as much as 2.5 times (fig. 2), a mismatch caused by two factors: high levels of uranium mined for military purposes, and a slower growth in the nuclear power industry than had been expected (NEA–IAEA, 2010).

Uranium Supply Worldwide

Primary Sources

The long-term operation of nuclear powerplants and the expansion of a nation's capacity for producing nuclear power depend on the development of uranium from primary sources. Yellowcake, or uranium oxide (U_3O_8), is the primary product of uranium mining, and the price of yellowcake hinges on world demand. Increases in uranium price encourage exploration for primary resources, thereby increasing supply. Mineability of an individual deposit is influenced by the delineation of identified RAR, the duration of the permitting process, the costs to mine and mill the product, the construction of infrastructure, and the ability of mine owners to raise capital to finance mining projects. Current estimates show the lag time from discovery to production ranges from 15 to 20 years (Vance, 2005; Boytsov, 2010) (fig. 3). When researchers attempt to determine world uranium supplies that will be available in the future, the length of this lag time makes it necessary to look at projects that are in early stages of development.

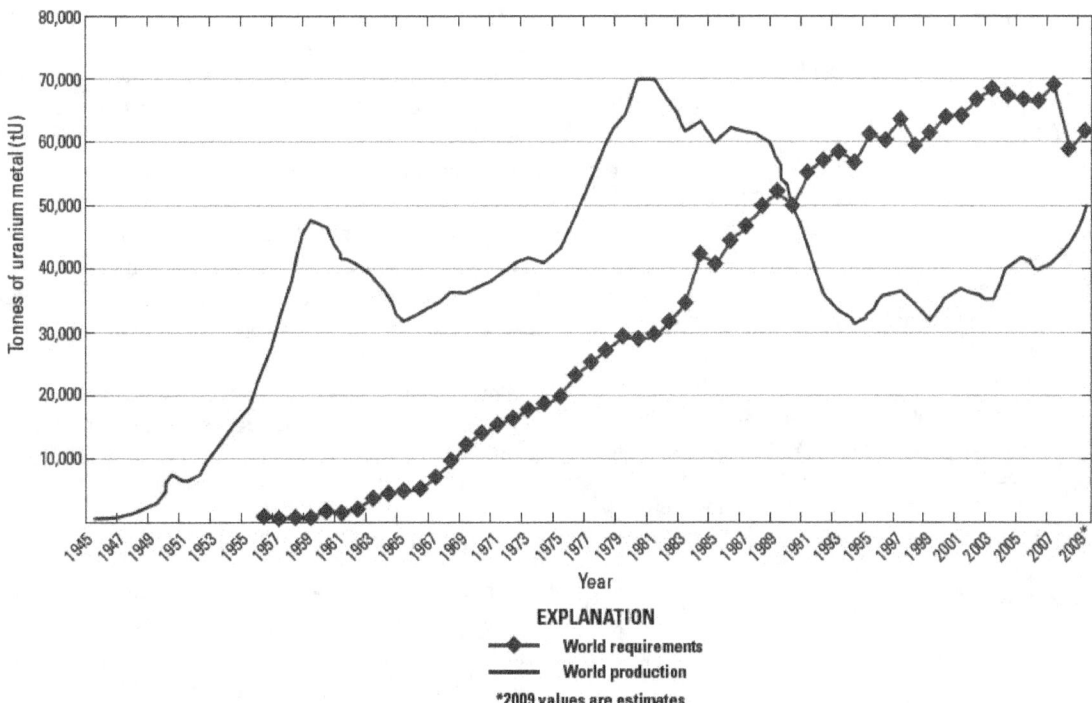

Figure 2.   Historic uranium production and nuclear powerplant requirements, 1945–2009. From NEA–IAEA (2010), reproduced with permission.

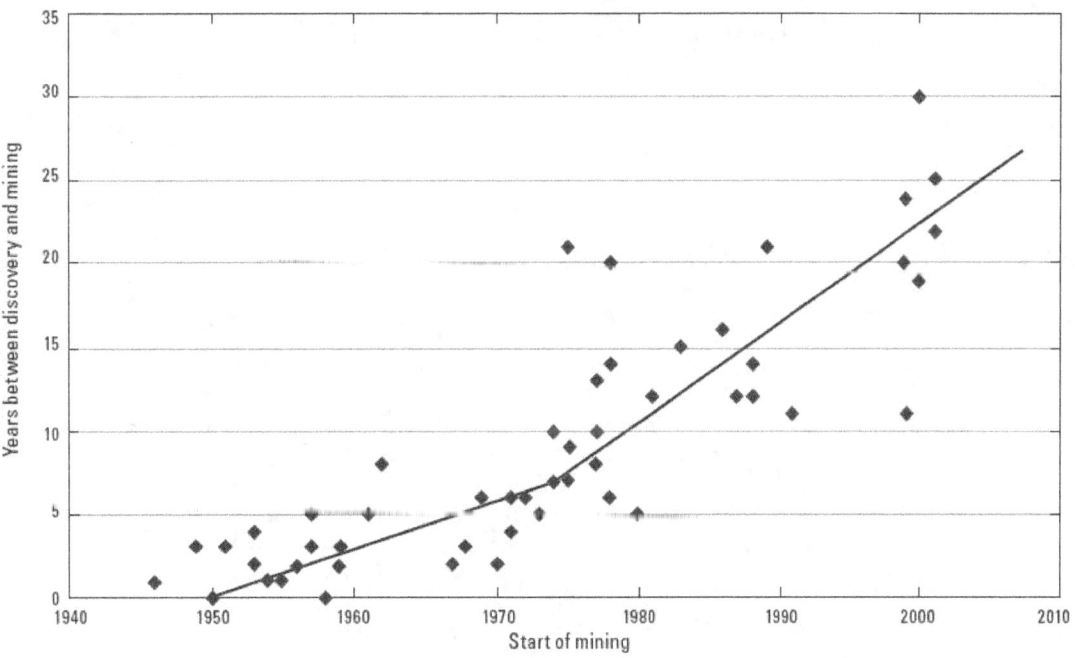

Figure 3.   Elapsed time between discovery of a uranium mine and the start of mining, for all mining methods. From NEA–IAEA (2010), reproduced with permission.

Identification of Primary Sources

The development of uranium supply begins when new uranium-producing properties are identified. An analysis of expenditures for exploration provides a measure of the effort being expended to identify new resources and to bring supply online. Since 1975, exploration expenditures increased rapidly until 1980, then increased at a more gradual rate through 2006, when a rise in the price of uranium preceded a rapid infusion of exploration expenditures. Increases in RAR followed the increase in exploration activities as measured by exploration expenditures (fig. 4).

Most uranium mining districts were originally identified by mineralization of uranium that was exposed at the surface, producing a geochemical and (or) geophysical anomaly. Studying the geology of these exposed deposits enabled geologists to identify concealed deposits in the same local environment. Identifying deposits having no such surface expression may in the future require more resources, including time, to delineate these targets than did deposits discovered in the past.

At the reconnaissance scale, conventional geophysical and geochemical techniques have been used to identifying concealed uranium deposits with mixed success. Radon, a uranium-decay product, has been analyzed on surfaces above potential deposits, but because this decay product is short lived and relatively mobile anomalies do not always directly identify mineralization. Geochemical analysis of groundwater has been used successfully in locating deposits which do

not crop out. Some geophysical surveys effectively target units that commonly contain uranium mineralization, such as conductive shale units in the Athabasca basin (Saskatchewan, Canada) that can then be explored by drill testing. Recently, variable time-domain electromagnetic techniques have identified uranium breccia-type deposits in Arizona (Spiering and others, 2009).

Deposit-scale geophysical techniques have proved more successful. It is now possible to directly measure U^{235} using Prompt Fission Neutron technology, which is being used in lieu of indirect measurement of U^{235} by interpretation of gamma profiles. However, at the reconnaissance scale this technique is of limited use in identifying concealed deposits. If better techniques to identify concealed deposits are developed, it is likely that more uranium supply could be identified.

Costs of Uranium as a Fuel

The cost of uranium fuel for generating electricity is low when compared to the costs of other types of fuel. Although finding accurate figures is difficult, the price of yellowcake is estimated to contribute only about 25 percent to the total cost of nuclear fuel, the rest being attributable to processing (conversion, enrichment, and fabrication). At 60 U.S. dollars (USD) per pound of uranium oxide (U_3O_8), (equivalent to USD 155 per kilogram contained uranium metal (USD 155/kgU)), nuclear fuel costs less than 0.7 cents

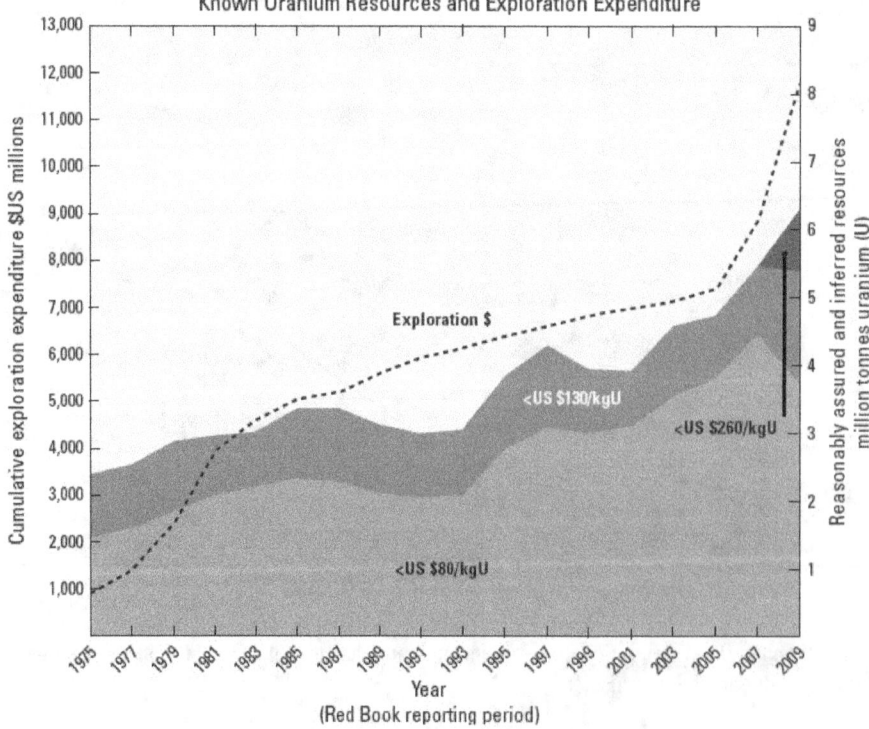

Figure 4.   Uranium exploration expenditures and resources identified during 1975–2009. From World Nuclear Association (2011b), reproduced with permission.

per kilowatt-hour (kWh) of electricity, or on average about 4 to 6 percent of the retail price of electricity (MIT Energy Initiative, 2011). This low relative fuel cost makes it easier for utilities to absorb increases in uranium price than to absorb price increases for more costly fuels that generate electricity. MIT estimates that utilities can absorb the uranium costs of USD 300 to USD 400/kgU for light water reactors. This would increase lifetime-levelized costs (busbar costs) for nuclear reactors by 8 to 12 percent (MIT Energy Initiative, 2011). If uranium prices doubled, an estimated 479,000 tU of RAR of uranium in deposits that are now subeconomic would potentially become economic.

Contractual Categories for Purchasing Uranium

Worldwide uranium purchases fall into two categories: spot purchases (delivery within one year), and contracts (medium- and long-term delivery). These prices have traditionally tracked each other fairly closely, with the exception of the time period of 2006 to 2009 when market forces caused the prices to decouple (fig. 5). While waiting for sale or delivery, U_3O_8 (uranium as yellowcake) can be held only at producers' sites or at conversion sites.

Long-term contracts are those in which utilities contract with a supplier, most commonly a corporation owning an active mine, to supply their uranium needs for generating electricity. These contracts are typically at a fixed price, with provisions for fluctuations in market price and demand, and

they run for many years. The duration of long-term contracts depends upon where the buyer is physically situated. In the United States, contracts typically run for 5 years; in Europe, 10 years; and in Japan, typically 15 years. For price indicators, the industry relies on market research because these contracts are generally not publicly available; the exception being contracts in European Union countries which are reviewed by the EURATOM Supply Agency. The Ux Consulting Company LLC. (*http://www.uxc.com*) ,TradeTech (*http://www.uranium.info/*) and the Euratom Supply Agency (*http://ec.europa.eu/euratom/*) all track uranium prices.

The short-term "spot" price (available for delivery in a short time frame (3–12 months)) of uranium is a smaller market in total volume. In 2011, the volume of uranium in the spot market was about 16,000 tU, or 20 percent of demand, and 30 percent of production, a ratio similar to that in the 1990s (TradeTech, 2011; Ux Consulting Company LLC., 2011). Uranium ends up on the spot market from smaller mines that cannot supply the quantities of uranium over timeframes that utilities require, or as a speculative product, with intermediaries buying uranium and holding it in hopes of receiving a higher price in due time. Uranium can also end up on the spot market in special circumstances, such as those of the U.S. Department of Energy, which is currently selling U.S. uranium stockpiles to help meet its costs of environmental cleanup at the Portsmouth, Ohio enrichment facility. Buyers on the spot market can be utilities, producers, or intermediaries, the uranium being either used in reactors or being resold.

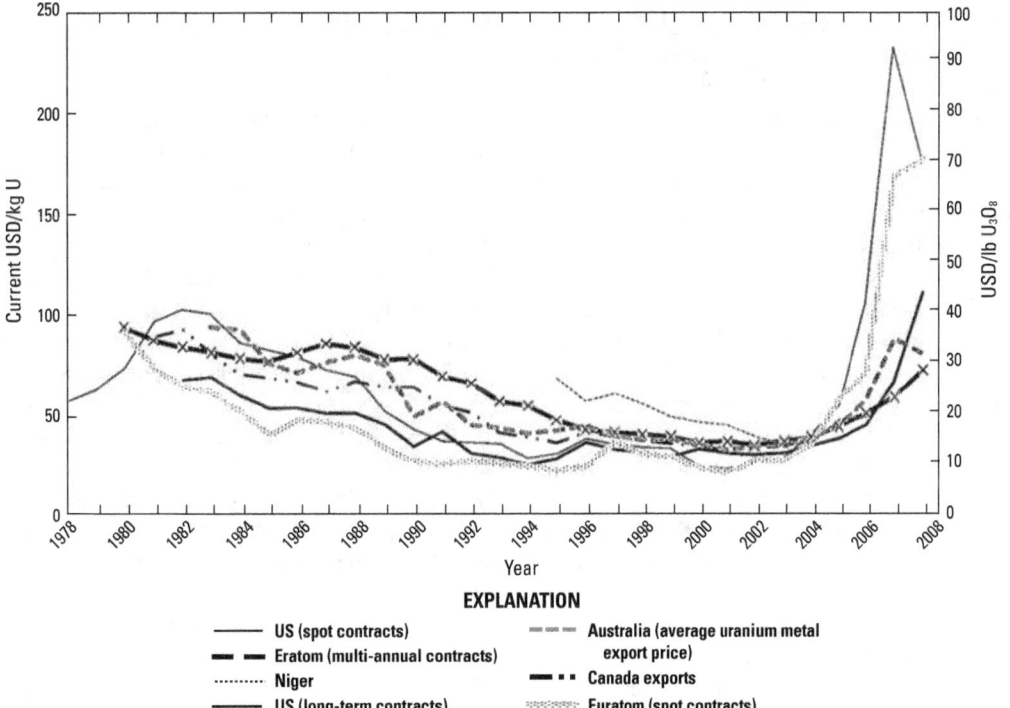

EXPLANATION

——— US (spot contracts)

— ■ — Eratom (multi-annual contracts)

·········· Niger

——— US (long-term contracts)

— — — Australia (average uranium metal export price)

— ■ ■ — Canada exports

░░░░░ Euratom (spot contracts)

Figure 5. Uranium prices in U.S. dollars per pound of uranium oxide (U_3O_8), by type of contract, by major buyers, 1978–2009. From NEA–IAEA (2010), reproduced with permission.

Producers, traders, and even utilities at times may all contribute to the spot market. Key traders are NUKEM BMbH (and NUKEM Inc.), Urangelsellschaft mbH, INTERNEXCO GmbH, Marubeni Corporation, Traxys North America LLC, ITOCHU Corporation, Nufcor International Ltd.(Goldman Sachs) and the Mitsui Corporation. Some producers, such as the Cameco Corporation (Cameco), also participate in the spot market. Brokers seek out uranium and place it for a commission, including American Fuel Resources, ICAP, MF Global, the New York Nuclear Corporation, and Evolution Markets. Hedge fund managers and investor funds have become attracted to the spot uranium market, running up prices to a recent high price of USD 136/lb. U_3O_8 (USD 353/kgU) in July 2007. This kind of investing decoupled spot and long-term prices to such an extent that the spot price now responds equally rapidly to perceived and to real threats to uranium supply. For this reason, short-term prices may not be the best indicator of the cost of nuclear fuel, which more closely tracks long-term contract prices.

Secondary Sources

Secondary sources are likely to become increasingly important for meeting uranium demand over the longer term. Projections to 2020 predict that the contribution from secondary supplies will shrink, while primary supply, mainly from mines in Africa, Kazakhstan, Australia and Canada, increases. (figs. 6, 7) (Ux Consulting Company LLC, 2010).

The more important secondary sources, the uranium that may be included in these resources and supply challenges are described below. Although secondary uranium sources are an important portion of the total world uranium supply, the quantity of uranium contained in these sources is difficult to quantify. Most countries do not report stockpiles of uranium, the concentration of uranium in depleted uranium tails is not well quantified, nor is information about the use of tails readily available, and the future disposition of Russian HEU is unknown.

Stockpiles.—From 1945 through 2008, NEA estimates that 2,415,000 tU were produced and that 1,840,000 tU were consumed, with the surplus production of 575,000 tU remaining in stockpiles (NEA–IAEA, 2010). However, the amount of this material that could become available to the market is not well known, since only limited information on the size of world stockpiles of uranium is publicly available.

HEU to LEU.—Programs that reduce HEU (highly enriched uranium) to LEU (low enriched uranium) ("HEU to LEU") are another secondary uranium source. In the United States, the Megatons to Megawatts program, a government–industry partnership in which Russian-origin HEU is downblended for use in nuclear power plants, is expected to end in 2013, reducing the supply of secondary uranium

by an estimated 9,200 tU/yr. To date, this program is estimated to have recycled more than 400 metric tons of HEU into 11,905 tU of LEU for use in U.S. nuclear powerplants (U.S. Enrichment Corporation, 2011).

Re-enriched Tails.—Depleted uranium tails are a byproduct of the uranium enrichment process. NEA (2010) estimates that 1,600,000 metric tonnes is contained in uranium tails worldwide, at an average concentration between 0.25 and 0.35 percent U. This grade is similar to uranium grades in mines that are economically extracting uranium from sandstone-hosted uranium deposits. NEA estimates that, if this entire inventory were re-enriched, 450,000 tU would be produced, the equivalent of more than 7 years of consumption at 2010 levels. However, this enrichment requires commercial capacity that is currently not available for enrichment and that only high uranium prices could sustain.

MOX and RepU.—Mixed oxide fuel (MOX) and reprocessed uranium (RepU) are expected to become increasingly important secondary sources of supply in the future. MOX and RepU originate from reprocessing spent nuclear fuel. Uranium and plutonium are recovered by reprocessing, and can then be used in nuclear power plants. The use of RepU fuel is tied to uranium costs; when mined uranium carries higher costs,, reprocessed fuel becomes more attractive. MOX fuel is widely used in reactors in Japan and Europe. Fifty reactors worldwide are licensed to use MOX fuel, although not all of them are using this fuel type (World Nuclear Association, 2011b).

Unconventional Resources

Uranium recoverable only as a mining byproduct is termed an "unconventional" uranium resource. The contribution of uranium from unconventional resources could be an important source in the future. Uranium in phosphate-rich rocks, in black shales, in lignite, and in seawater are considered unconventional resources.

Phosphates have historically been a source of market supply of uranium and are a potential source of uranium in the future. Prior to December 2011 an estimated 57,863 tU was produced from phosphate deposits in Kazakhstan, the United States and Morocco (NEA–IAEA, 2010). Production costs higher than the market value of uranium have slowed development of this supply. However, recent technical innovations that hold the promise of more cost-effective production of uranium from phosphate deposits have prompted industry investment into developing this resource (Jones and others, 2009; World Nuclear News, 2007). Cameco invested in and is testing the effectiveness of the PhosEnergy process developed to extract uranium from phosphate rock (Ux Consulting Company LLC, 2010). Uranium in phosphate rocks is estimated to contain a resource greater than 7.9 million tU with

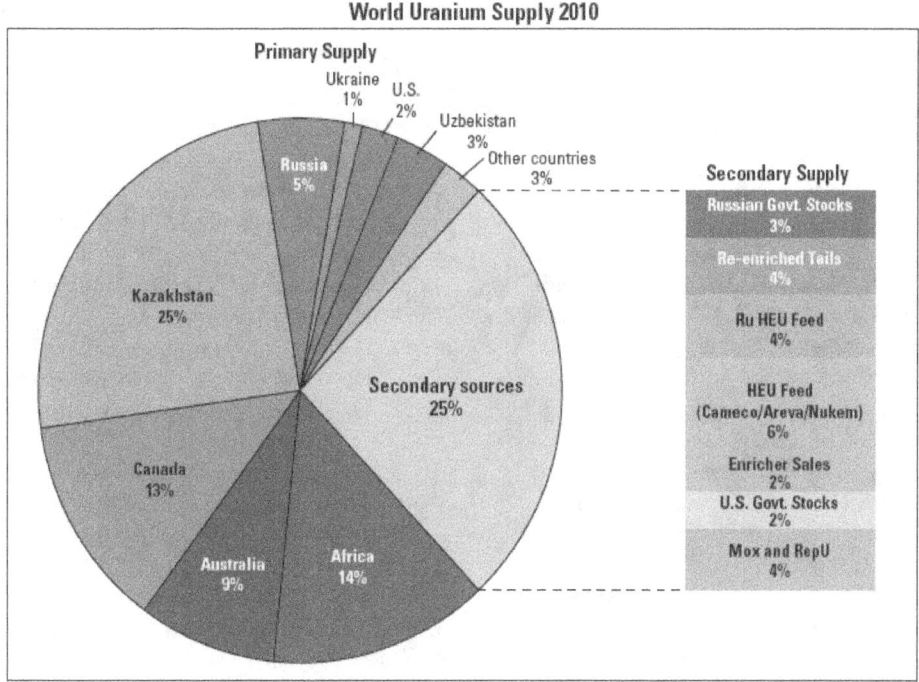

Figure 6. World uranium supply distribution, 2010. Data from Ux Consulting Company LLC (2010).

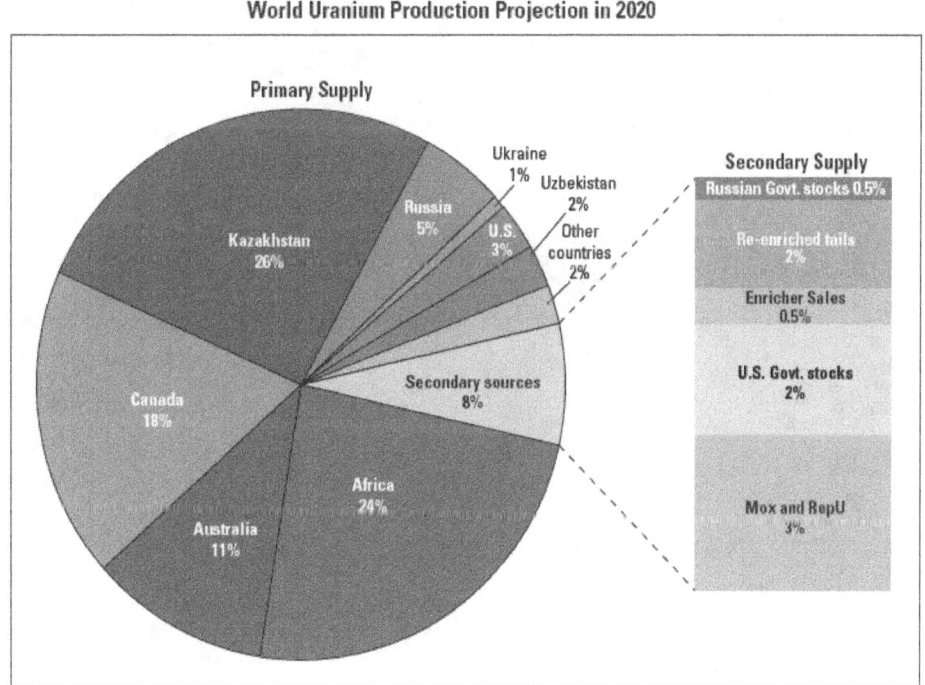

Figure 7. Contribution of primary and secondary uranium sources worldwide, projected to 2020. Data from Ux Consulting Company LLC (2010).

an average ore grade that ranges from 0.006 percent U to 0.053 percent U (IAEA, 2010). Uranium rich phosphate deposits are identified in Morocco, the United States (Florida and Idaho), Jordan, Egypt, Iran, Iraq, Mexico, Kazakhstan, Sweden, Syria, Israel, Brazil, Finland, and Greece (IAEA, 2010). Pilot projects exploring uranium extraction from phosphate deposits are underway in the United States, Brazil, and Jordan.

An estimated 1.3 million tU worldwide is contained in black shale and in lignite deposits (IAEA, 2010). The uranium in these deposits is low-grade (0.01 to 0.17 percent U) and requires production of another commodity to support extraction of the uranium (IAEA, 2010). The largest uranium-rich black shale deposits are in Sweden and Germany, with deposits also identified in Uzbekistan, Korea, China, Canada, Poland, Turkmenistan, Finland, Uzbekistan, Poland, and France. Uraniferous lignites are identified in South Africa, Kazakhstan, Russia, Spain, the United States, Australia, Greece, Germany, Kyrgyzstan, and the Czech Republic. Uranium has been mined from black shales in the past in Sweden. The Talvivaara polymetallic black shale deposit in Finland, currently being mined for nickel and zinc, also contains an estimated 17,110 tU. Cameco is financing the construction of a circuit to recover uranium from this resource, targeting a production rate of 350 tU/yr (Ux Consulting Company LLC, 2010).

Research to develop cost-effective techniques to recover uranium from seawater has been carried out in Germany, Italy, Japan, the United States, and the United Kingdom. The current focus of research is the development of specialized polymer braids moored on the ocean floor. Recovery costs for a large-scale system that would recover 1200 tU/yr are estimated to be about USD 700/kgU (Hisatani, 2010; NEA–IAEA, 2010), which would be uneconomic at current and anticipated prices of uranium.

Current World and U.S. Production

World uranium production in 2010 was 53,663 tU, up from 50,772 tU in 2009 (World Nuclear Association, 2009; NEA–IAEA, 2010). Global production has increased gradually since the early 1990s, after steadily declining for 13 years (1980–93). Six countries currently dominate world production and are expected to produce 83 percent of uranium concentrate during the 10 years until 2020: Kazakhstan, Canada, Australia, Namibia, Russia, and Niger (Ux Consulting Company LLC, 2010). The three largest producers—Kazakhstan, Canada, and Australia—account for 63 percent of the world production of uranium concentrate (World Nuclear Association, 2009). Kazakhstan dramatically increased uranium production from 2,022 tU in 2001 to 19,450 tU in 2011, although projections indicate peak production capacity may have been reached (World Nuclear Association, 2011b).

The United States currently produces about 3 percent of the world uranium concentrate (NEA–IAEA, 2010), with 4.2 million pounds U_3O_8 (1,629 tU) in 2010 from one mill (White Mesa Mill) and from four *in-situ* leaching (ISL) plants (Alta Mesa Project, Crow Butte Operation, La Palangana, and Smith Ranch–Highland Operation) (Energy Information Administration, 2010a).

Domestic and world uranium production and exploration has historically responded to market conditions. For example, U.S. production of uranium concentrate peaked in 1980 at more than 43 million pounds of U_3O_8 (16,810 tU), as a prolonged period of rising prices and intensive exploration ended (fig. 8). By 1980, production exceeded reactor requirements, creating a surplus of uranium. During 1981–2003, domestic uranium production declined to a low of 2 million pounds of U_3O_8 (769 tU), coinciding with a nearly 20-year period of falling prices (1981–2000). Prices began to rise in 2000, with significant increases during 2003–2007. Increases in the spot price of

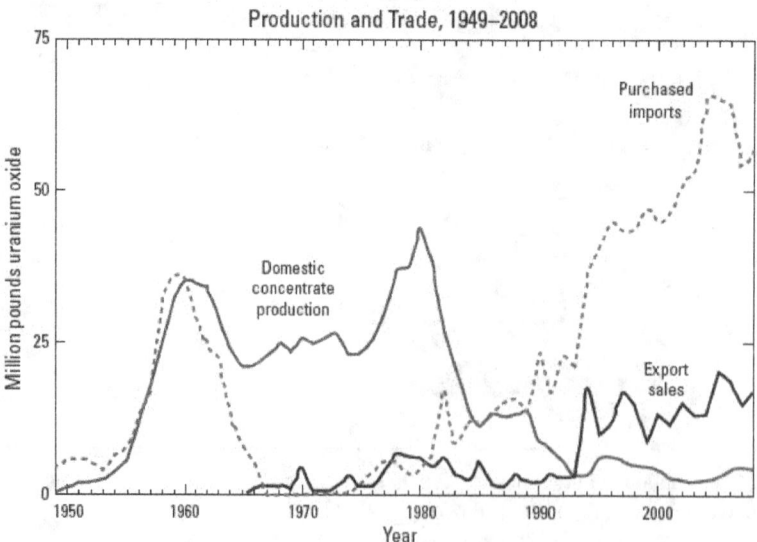

Figure 8. U.S. production, purchases, and sales of uranium, 1949–2008 (Energy Information Administration, 2010b).

uranium during 2000–2007 were attributed to a combination of market factors: the increasing prospects for nuclear power-plant construction, declining inventories, temporary difficulties at existing and developing mines and mills, and the entry of speculators into the uranium market (Nuclear Energy Agency, 2008). U.S. production also steadily increased from less than 2 million pounds (769 tU) in 2003 to more than 4.5 million pounds of U_3O_8 (1,730 tU) in 2007, following the increase in spot prices. Uranium prices reached a high of USD 136/lbU_3O_8 (USD 353/kgU) in spring 2007, followed by a drop to prices ranging from USD 40 to USD 55 during the three years 2008–2011, as a rapid expansion of production in Kazakhstan. Other market factors, such as the global financial crisis, added obstacles to financing uranium exploration, mine development, and construction of nuclear power plants, all of which contrib-uted to the "cooling" of the uranium market. Prices surged again during the last quarter of 2010 and in early 2011, in response to China's announced plans for and its moves to secure uranium contracts for large planned increases in nuclear power. Sus-tained higher prices would most likely stimulate uranium explo-ration and production, although permitting of new and expanded mines continues to be challenging.

Uranium Demand Worldwide

The only uranium requirements that this analysis con-siders are those necessary for the generation of electricity by civilian nuclear powerplants. Military and other government requirements are not included. Growth in world generation of electricity has outpaced growth in total consumption of energy during the 20-year period, 1991–2011; this trend is expected to continue for generating electric power through 2035 (Energy Information Administration, 2010b). Nuclear power accounts for about 14 percent of worldwide and 20 percent of domes-tic U.S. generation of electricity (Energy Information Administration, 2010b). Although electricity generated by nuclear power is expected to increase by about 2 percent a year for the 25 years through 2035, the relative contribution of nuclear energy to the generation of electricity is expected to stay the same (Energy Information Administration, 2010b) (fig. 9).

Current World Uranium Demand

In 2010, the world demand for uranium to power com-mercial reactors for electricity generation was 68,646 tU, as measured from acquisitions of uranium resources (World Nuclear Association, 2011b). The total acquired uranium for nuclear power is not an exact measure of the amount of uranium actually loaded into reactors; it may be higher or lower than the amount used for power generation, depending on the amount used from inventories (World Nuclear Association, 2011a).

Current U.S. Uranium Demand

In 2009, owners and operators of U.S. civilian nuclear reactors purchased a total of 50 million pounds of U_3O_8 (19,232 tU) (Energy Information Administration, 2009). Fol-lowing recent trends, most uranium purchased in the United States in 2009 (86 percent) originated from foreign produc-ers, while 14 percent originated from U.S. mining operations (fig. 10). In 2009, uranium in U.S. nuclear reactors originated

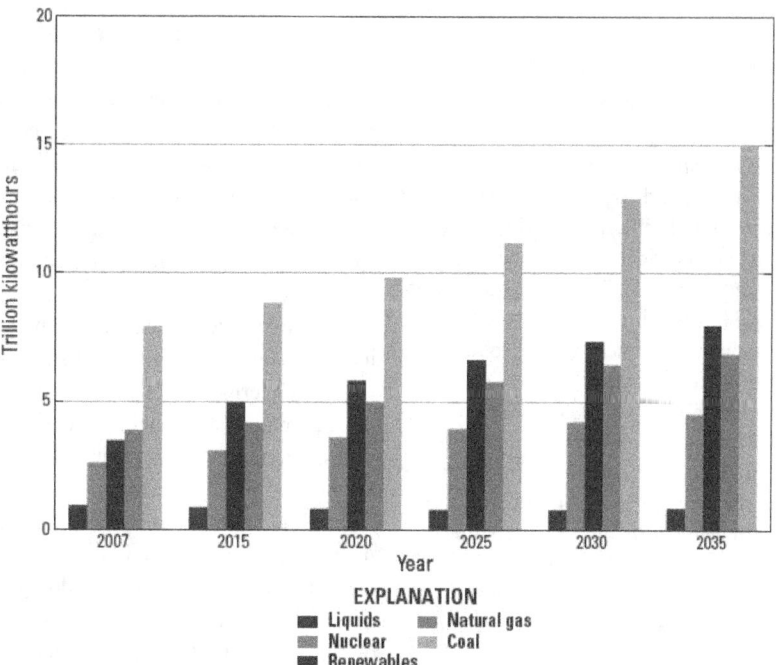

Figure 9.   Net generation of electricity worldwide, in trillions of kilowatt hours, by all fuels, 2007–2035 (Energy Information Administration, 2010c).

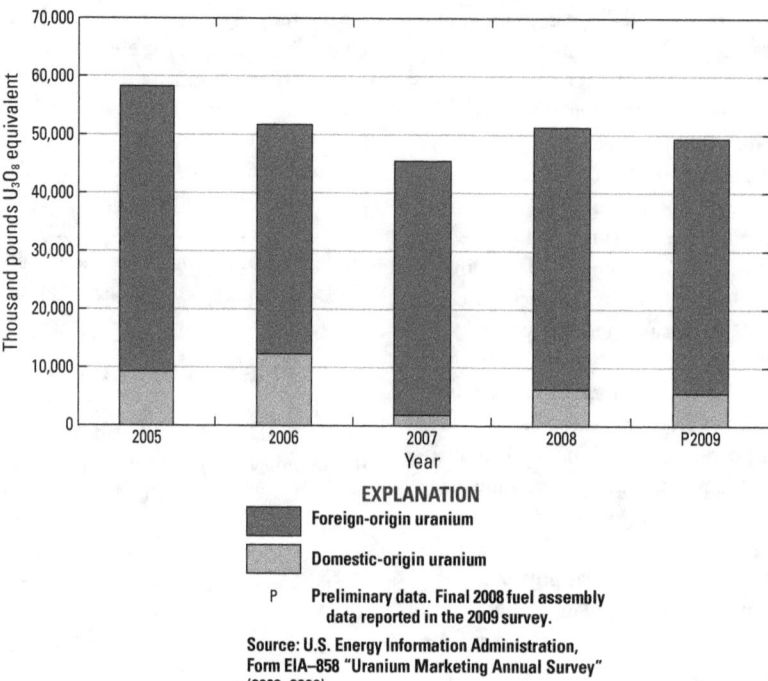

Figure 10. Uranium in fuel assemblies, in thousand pounds of uranium oxide equivalent loaded into U.S. civilian nuclear power reactors, by originating source, by year, 2005–2009 (Energy Information Administration, 2010d).

from Australia and Canada (40 percent); Kazakhstan, Russia, and Uzbekistan (29 percent); and Brazil, Czech Republic, Namibia, Niger, and South Africa (17 percent) (Energy Information Administration, 2010d) (fig. 11).

Current U.S. Uranium Inventory

The commercial inventory of uranium owned by U.S. civilian nuclear powerplant owners and operators totaled 84 million pounds U_3O_8 (32,310 tU) by 2009 year end. Commercial inventory includes ownership of uranium in various stages of the fuel cycle at domestic and foreign facilities. The total U.S. commercial inventory including inventories owned by brokers, converters, enrichers, fabricators, producers, and traders, was 110 million pounds U_3O_8 (42,311 tU) at the end of 2009 (Energy Information Administration, 2010d). In addition to their existing inventories, owners of nuclear powerplants have contracts in place for uranium for which EIA collects data 10 years into the future. At the end of 2009, commercial plants had purchase contracts in place for a total of 261 million pounds of U_3O_8 (100,392 tU)) under purchase contracts during 2010–2019. The maximum anticipated market requirement, for commercial plant owners alone, during 2010–2019, totals 503.4 million pounds U_3O_8 (193,477 tU) (Energy Information Administration, 2010d) (fig. 12). Note that "market requirement" is not the same as "commercial reactor requirement," although the numbers are not significantly different (~50 million pounds U_3O_8, (19,232 tU) per year loaded into U.S. commercial reactors).

Projected Future Uranium Supply and Demand

Growth in Demand

Projections of future uranium demand depend on predictions of nuclear generating capacity, and on the type of reactors and fuel being used to generate electricity. EIA forecasts that electricity generation from nuclear power worldwide will increase from 2.6 trillion kilowatt-hours in 2007 to 4.5 trillion kilowatt-hours in 2035. Global concerns about greenhouse gases, rising fossil-fuel prices, the need for additional energy in developing countries and energy security support the development of additional nuclear capacity. However significant challenges and uncertainties remain, including unresolved issues of storage and disposal of nuclear waste, concerns about the safety of nuclear power, and the large capital costs associated with powerplant construction. These major concerns continue to prevent significant growth of nuclear power in many member countries of the OECD. Several nonmember countries, most notably China, are forging ahead with construction of new powerplants, and they maintain ambitious goals for adding significant new capacity during the 25 years to 2035. In the longer term, the expansion of the use of MOX and RepU fuels, and the development of Generation IV reactors, with their lower fuel requirements, will also influence demand.

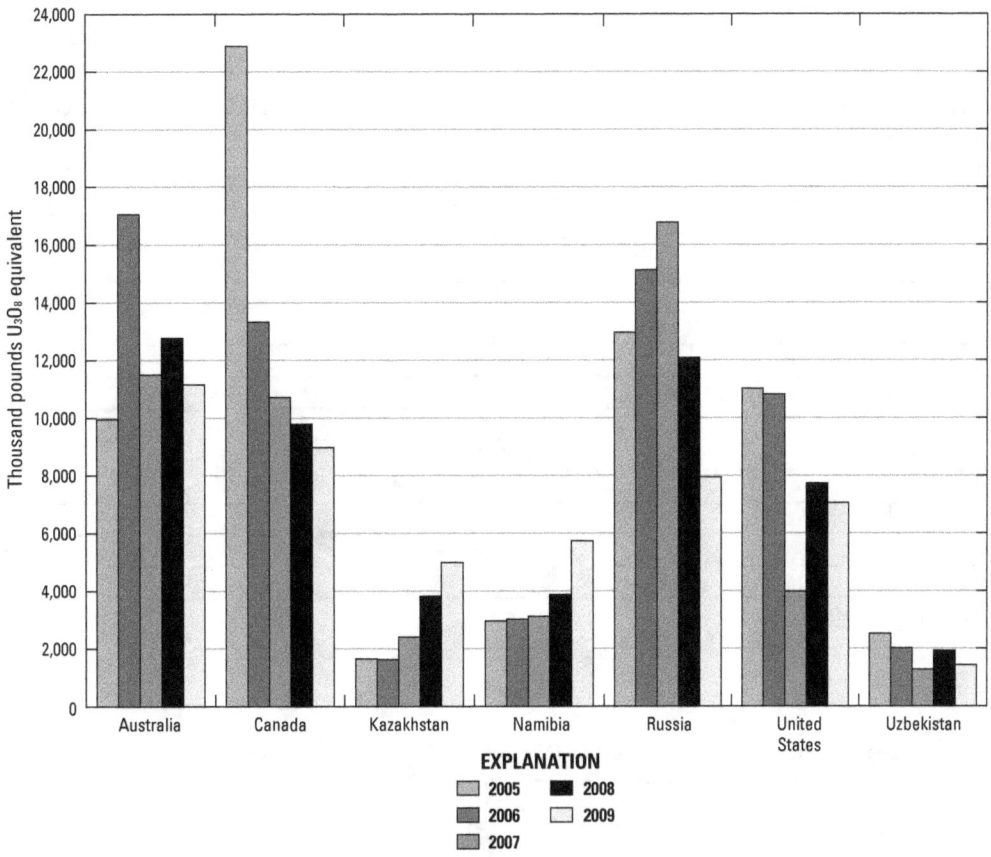

Figure 11.   Uranium purchased by owners and operators of U.S. civilian nuclear power reactors, by selected country of origin and by delivery year, 2005–2009 (Energy Information Administration, 2010d).

Short-term projections are made with a fair amount of certainty, because most near-term capacity is already in operation. Longer term forecasts, to 2035, 2050, and 2100, are subject to much greater uncertainties. The growth of nuclear capacity in a given country depends on economics, which are difficult to predict, and on legislation and regulations that are subject to change. In any case, most scenarios point to future growth. To account for the uncertainties, the following projections consider both low- and high-case scenarios.

Growth in Supply

In order to assess with some certainty the issues of uranium supply in the near future (~10 years), this study evaluates individual deposits and aggregates these results to determine longer-range trends. Beyond 10 years, development is much more uncertain. The NEA and the IAEA jointly prepare the biennial publication "Uranium 20XX: Resources, Production and Demand," also known as the " Red Book" for its distinctive red cover. For this report, published studies of long-term supply were critically examined, and resources from the "Red Book" were used to estimate long-term uranium supply (NEA–IAEA, 2010).

Tables 1–4 and appendix 1 provide detailed summaries of operating uranium mines and of properties that are likely to be producing uranium in the near future. No single published

source exists for the tonnage of remaining RAR within operating mines, and so this paper examined each deposit using best estimates. Sources for the report's information are presented in tables 1–4. Some countries, such as India and Iran, are not expected to produce uranium that will be sold on the open market, but this report evaluated the RAR within these countries and included those data in its totals for comparing worldwide uranium supply to demand.

Production capacity is reported for mines where available. No mine operates at its maximum-rated capacity for the entire mine life, and so readers should consider the stated capacity to be a guideline, useful in terms of estimating short-term supply only. A production center can process ore from several mines, as the White Mesa Mill in Utah processes ore from the Arizona One mine and from Colorado's Pandora and Daneros mines. Alternatively, a production center may represent an ISL mine that produces yellowcake from each individual mine as a final product without offsite milling. Production facilities, usually uranium mill sites for conventional mining, and the mines that supply ore to these facilities, were cross-checked to avoid over reporting of capacity. In some cases, data were unclear when describing which production facilities were supplied by which mines. As well, reports of RAR may not be accompanied by data on the proposed capacity for these mines.

2009 Uranium Marketing Annual Report
Release Date: August 18, 2010
Next Release Date: May 2011

Maximum Anticipated Uranium Market Requirements of Owners and Operators of U.S. Civilian Nuclear Power Reactors, 2010–2019, as of December 31, 2009
(Thousand Pounds U$_3$O$_8$ Equivalent)

Year	Maximum Under Purchase Contracts	Unfilled Market Requirements	Maximum Anticipated Market Requirements	Enrichment Feed Deliveries
2010	40,739	4,425	45,164	47,567
2011	39,836	5,688	45,523	49,621
2012	36,296	15,342	51,638	52,712
2013	34,846	16,988	51,834	55,712
2014	31,025	17,725	48,749	48,746
2015	25,691	26,892	52,583	52,020
2016	20,004	32,238	52,242	53,686
2017	14,722	34,113	48,834	51,217
2018	10,863	44,661	55,525	56,379
2019	6,961	44,374	51,334	51,575
Total	**260,982**	**242,444**	**503,426**	**519,237**

Note: Totals may not equal sum of components because of independent rounding.

Source: U.S. Energy Information Administration: Form EIA-858 "Uranium Marketing Annual Survey" (2009).

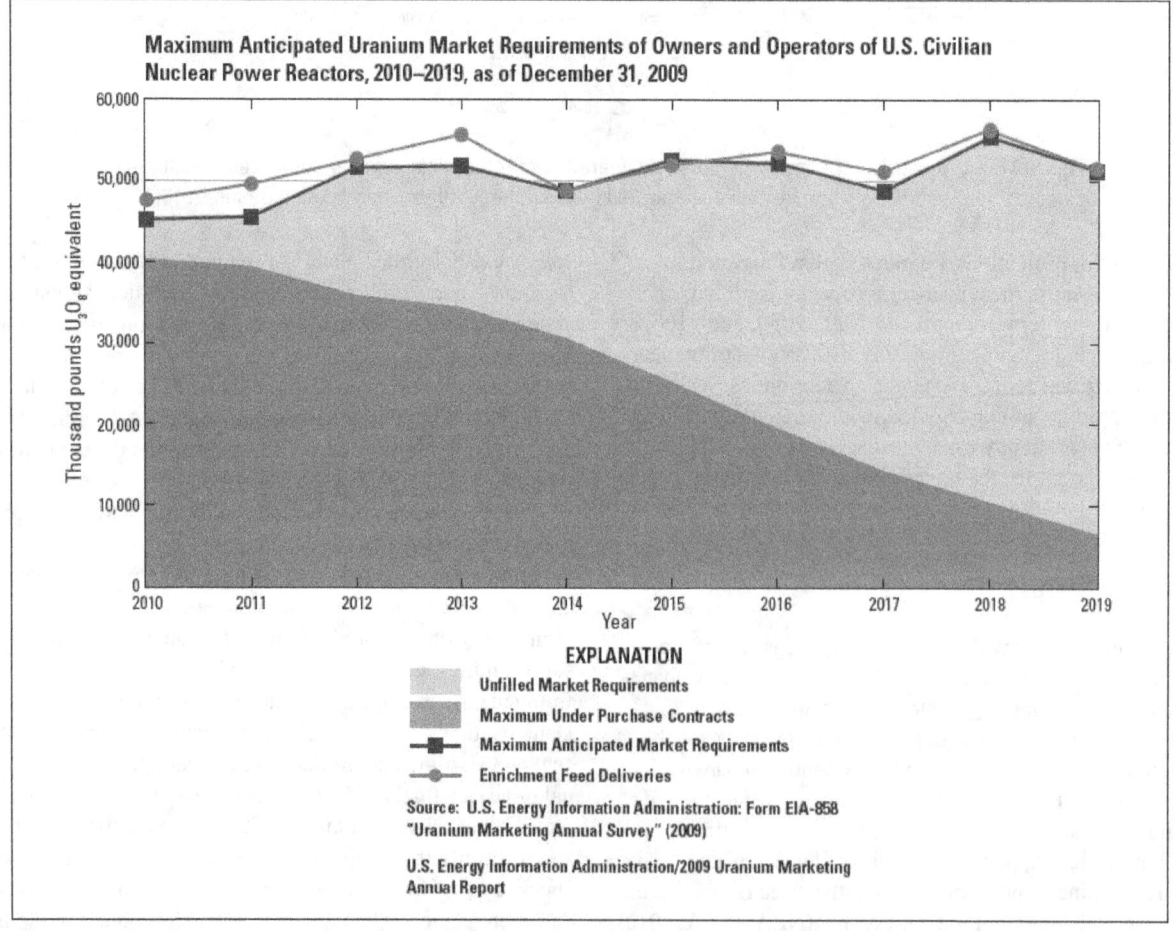

Figure 12.   Maximum anticipated uranium market requirements of owners and operators of U.S. civilian nuclear power reactors, 2010–2019, as of December 31, 2009 (Energy Information Administration, 2010d).

Beyond fact-checking of RAR, of capacity, and of mine life, country narratives explore potential interruptions to supply based on technical, economic, or political challenges for individual deposits or for political provinces (appendix 1). In these narratives the term "in the near future" describes a development scenario that is ten to fifteen years into the future. This timeframe roughly corresponds to the time it takes to develop a uranium deposit into a producing mine and, in some cases, to develop mining regulations within countries where such regulations do not exist.

Identified Resources

Short-term Supply

An analysis of producing mines (operating mines and developing mines) completed for this report identifies 2.7 million tU of *in-situ* RAR (2.1 million tU estimated recoverable) in 64 operating and in 77 developing mines worldwide (tables 1–4). The operating mines/production centers report a total of 1.4 million tU of *in-situ* RAR (table 3), approximately 1 million tU of which are estimated to be recoverable (table 2, Developing mines report 1.3 million tU *in-situ* RAR (1.1 million tU estimated recoverable), which have a reasonable chance of producing uranium within 5 to 10 years (table 3). This report applies mining and milling to *in-situ* resources, following NEA/IAEA guidelines (NEA–IAEA, 2010). In ambiguous mining and resource scenarios, this report used conservative mining and milling losses, and so the resulting recoverable resources should be considered minimums. Because actual recovery is unknown until mining is complete, these are rough estimates to be used for planning purposes only. Recovery ranges from 65 percent to 80 percent, depending on the type of mining used to extract ore and on the specific metallurgical factors that are used to produce uranium from this ore.

The "2009 Red Book" reported 3.5 million tU of economic recoverable RAR worldwide (extractable for less than USD 130/kgU) (fig. 13, table 5) (NEA–IAEA, 2010). It was expected that the RAR in the development pipeline as estimated from this analysis would represent a subset of NEA's total reported economic reserve. The reserve base that is currently coming online represents 60 percent of RAR reported in the NEA–IAEA cost categories "less than USD 130/kgU" (fig. 13), which are economic at current uranium prices.

At 2010 rates of demand (68,646 tU; World Nuclear Association, 2011b), uranium in operating mines, or those that will be coming online in the near future as calculated in this report, would fuel the world nuclear power reactor fleet for about 30 years, depending on the percentage of actual uranium recovery during mining. However, this rate does not take into account future demand from the 61 reactors under construction, from 158 reactors planned or on order, or from 326 proposed reactors (World Nuclear Association, 2011b).

Reserve Distribution by Projected Mining Method

Conventional operating mines contain 68 percent of *in-situ* RAR, and 63 percent of the identified nominal capacity for uranium production, whereas ISL mines contain 32 percent of RAR and contribute 37 percent of reported capacity (table 6). ISL mines currently operate in Kazakhstan, the United States, Australia, China, Uzbekistan, and Russia (with one ISL mine planned for Pakistan). The share of future ISL capacity measured from advanced-stage properties (those expected to produce within the five years 2011–2015) is expected to fall to 20 percent of the total uranium being produced, and resources attributable to ISL in upcoming mines is a much lower 16 percent. Conventional mines expected to come online in the next 5 to 10 years (2011–2021) are estimated to contribute 84 percent of the resources, and to provide 80 percent of the capacity to future production. However, many properties in development do not report proposed capacity. Therefore future production estimates should be considered as minimums, and the ratios of future ISL to conventional mined uranium should be considered as supply estimates only. Note that the category of "developing conventional mines" includes proposed mines for which a production method is undetermined, byproduct production, and mining of dumps, in addition to open pit and underground operations.

Top 10 Producers

The top 10 producing mines in 2009 contributed about 30,600 tU, or 62 percent of world production (table 1). Forty-four percent of the world's uranium resources in operating mines is contained in these ten mines. However, production is not expected to continue at current capacity from many of the top ten. Production from Ranger, the second largest producer in 2009, is likely to decline as the mine depletes known high-grade resources and moves to develop lower-grade ores on leach piles not originally processed, and deeper targets. Olympic Dam, the sixth largest producer, will continue to produce only if prices for copper and gold, the primary commodities, remain high, and if capital can be raised for a significant expansion. Ore at Rabbit Lake/Eagle Point, the eighth largest producer, is nearly exhausted.

Table 1. Top ten uranium producing mines in 2009: production, reasonably assured resourecs, and mine life.

[tU, metric tonnes]

Mine	Location	Majority owner	2009 Production (tU)	Percent of world total production in 2009	Reasonably assured resource (tU)	Percent of world reserves (2009)	Average ore grade (% U_3O_8)	Mine type	Expected mine life	Information source
McArthur River	Canada	Cameco	7,400	14.6%	128,900	7.8%	19.5	Underground—Mill	2030	World Nuclear Association, 2011a
Ranger	Australia	Rio Tinto	4,423	9.0%	28,832	2.0%	0.135	Open Pit—Mill; Expansion includes underground and heap leach	2014	McKay and Carson, 2010; NEA–IAEA, 2010
Rossing	Namibia	Rio Tinto	3,574	7.0%	17,007	3.5%	0.031	Open Pit—Tank Heap Leach	2021	Ux Consulting Company LLC, 2010
Myunkum/Moinkum (Katco)	Kazakhstan	Areva	3,250	6.0%	24,131	2.4%	0.074	ISL	2039	Ux Consulting Company LLC, 2010
Streltsovskoye (Priargunsky)	Russia	ARMZ	3,003	6.0%	118,340	7.2%	0.12–0.26	Underground—Heap Leach	Unknown	NEA/OECD–IAEA, 2010
Olympic Dam	Australia	BHP Billiton	2,981	6.0%	295,000	18.0%	0.059	Underground—Primary commodity copper/gold	2063	McKay and Carson, 2010; NEA–IAEA, 2010
Arlit	Niger	Areva	1,808	4.0%	23,171	1.1%	0.22–29	Open Pit—Mill	Unknown	UxConsulting, 2010; NEA/OECD–IAEA, 2010
Rabbit Lake	Canada	Cameco	1,400	3.0%	8,200	0.5%	0.75–0.89	Underground—Mill	2015	World Nuclear Association, 2011a
Akouta	Niger	Areva	1,435	3.0%	24,670	1.6%	0.34	Underground—Mill	Unknown	Ux Consulting Company LLC, 2010; NEA/OECD–IAEA, 2010
McClean Lake	Canada	Areva	1,400	3.0%	1,031	0.1%	0.68	Underground—Mill	Unknown	World Nuclear Association, 2011a
Total			30,674	61.6%	669,282	44.2%				

Table 2. World's largest deposits in stages of operation, development, feasibility or prefeasibility, 2010.

[tU, metric tonnes]

Mine	Location	Majority owner	2009 Uranium production (tU)	Reasonably assured resource (tU)	Status and mining method	Development time line	Geologic deposit type	Information source
Olympic Dam	Australia	BHP Billiton	2,981	295,000	Operating—(Underground/mill—(Primary commodity copper/gold)	Planning open pit expansion, will extend operations to 2063	Poly-metallic breccia	NEA–IAEA, 2010; McKay and Carson, 2010
Imouraren	Niger	Areva	0	183,520	Feasibility—(open pit	Production planned from 2013–2049	Sandstone	NEA–IAEA, 2010; Ux Consulting Company LLC, 2010
McArthur River	Australia	Cameco	7,340	128,900	Operating—(Underground/mill	Production planned to 2030	Unconformity	Ux Consulting Company LLC, 2010; World Nuclear Association, 2011b
Streltsovskoye	Russia	ARMZ	3,003	118,341	Operating—Underground/Surface heap leach	Unknown	Volcanic	NEA–IAEA, 2010; Ux Consulting Company LLC, 2010
Novokonstantinovskoe	Ukraine	VostGOK	0	93,630	Planned—Underground	Production by 2011	Metasomatite	NEA–IAEA, 2010; Ux Consulting Company LLC, 2010
Cigar Lake	Canada	Cameco	0	80,500	Planned—Underground	Production planned to begin in 2013	Unconformity	NEA–IAEA, 2010; Ux Consulting Company LLC, 2010
Uzbekistan All regions (Centres 1,2,3)	Uzbekistan	Navoi Mining and Metallurgy Combinat	NA	76,000	Operating—ISL	Unknown	Sandstone	NEA–IAEA, 2005, 2010
Elkon	Russia	ARMZ	0	71,300	Planned—Underground	Production to begin in 2016	Metasomatite	Ux Consulting Company LLC, 2010
Itataia-Santa Quiteria District	Brazil	Industrias Nucleares do Brazil	0	67,240	Feasibility—open pit	Production planned by 2013	Phosphorite	NEA–IAEA, 2010; Ux Consulting Company LLC, 2010
Marencia	Namibia	Marencia Energy Ltd.	0	62,856	Feasibility	13 year mine life	Surficial	Marencia Energy Limited, 2011; NEA–IAEA, 2010; Ux Consulting Company LLC, 2010
Langer Heinrich	Namibia	Paladin Energy	1,115	60,830	Operating—Open Pit	Operating to 2026 Expected to open in 2012	Surficial	NEA–IAEA, 2010; Ux Consulting Company LLC, 2010
Dominion Reef	South Africa	Shiva Uranium	0	55,753	Standby—Underground		Quartz-pebble conglomerate	NECSA, 2010
Inkai	Kazakhstan	Cameco	350	51,808	ISL	Operating to 2032	Sandstone	NEA–IAEA, 2010; Ux Consulting Company LLC, 2010
Kiggavik	Canada	Areva	0	51,574	Advanced Exploration—Permitting	Expected to begin production 2020	Unconformity	Ux Consulting Company LLC, 2010; World Nuclear Association, 2011b
Rossing	Namibia	Rio Tinto	3,574	50,657	Operating—Open pit/tank leach; expansion—heap leach planned	Production planned to 2021	Granite-hosted	Ux Consulting Company LLC, 2010
Yeeleerie	Australia	BHP Billiton	0	44,077	Planned—Open pit	20 to 40 years of production	Surficial	Ux Consulting Company LLC, 2010; BHP Billiton, 2011
Trekkopie	Namibia	Areva	0	42,243	Feasibilty—Open pit	Production by 2016 depending on uranium prices—10 years of production	Calcrete-type	NEA–IAEA, 2010; Ux Consulting Company LLC, 2010

Table 3.   Operating uranium mines and their remaining reasonably assured resources, 2010.

[tU, metric tonnes; tU/yr, metric tonnes per year; ISL, *in-situ* leach. *In-situ* resources reported, when 2009 Redbook cited information was for production or other descriptive information or where there was no other source of resource data]

Mine/ production center	Country	Remaining *in-situ* reserve tU (proven and probable economic)	Grade %U	Ownership	End of mine life
Beverley	Australia	12,192	0.229	Heathgate Resources Pty. Ltd., General Atomics USA	NA
Olympic Dam	Australia	295,000	0.059	BHP Billiton	2063
Ranger	Australia	28,832	0.117	Rio Tinto	2020
Lagoa Real-Caetite District	Brazil	12,700	0.300	Industrias Nucleares do Brazil	NA
McArthur River	Canada	128,900	15.72–26.33	Cameco Corporation	>2031
McClean Lake	Canada	1,031	0.530	Areva Resources Canada Inc.	NA
Rabbit Lake/Eagle Point	Canada	8,200	0.880	Cameco Corporation	2015
Dep.512/Yining	China	NA	NA	China National Nuclear Corporation	NA
Lantian Deposit	China	2,000	0.171	China National Nuclear Corporation	NA
Lianshanguan	China	1,000	0.340	China National Nuclear Corporation	NA
Qinglong district	China	8,000	NA	China National Nuclear Corporation	NA
Shaoguan	China	NA	NA	China National Nuclear Corporation	NA
Tengchong deposit	China	6,000	0.050	China National Nuclear Corporation	NA
Xiangshan district	China	29,000	0.100	China National Nuclear Corporation	NA
Xiazhuang district	China	12,000	NA	China National Nuclear Corporation	NA
Yili deposit	China	16,000	0.060	China National Nuclear Corporation	NA
Ziyuan	China	10,000	NA	China National Nuclear Corporation	NA
Rozna	Czech Republic	680	0.379	DIAMO state enterprise	2012 or until unprofitable
Straz	Czech Republic	1,320	0.030	DIAMO state enterprise	NA
Bagjata	India	2,106	0.047	Uranium Corporation of India	NA
Bhatin	India	2,200	0.050	Uranium Corporation of India	NA
Jaduguda	India	8,400	0.067	Uranium Corporation of India	NA
Narwapahar	India	11,500	0.050	Uranium Corporation of India	NA
Turamdih/Banduhurang	India	3,750	0.046	Uranium Corporation of India	NA
Gachin (Bandar-Abas)	Iran	100	0.200	Atomic Energy Organization of Iran	NA
Akdala	Kazakhstan	9,308	0.070	Uranium One Inc., KazAtomProm	NA
Budenovskoye 1-3-4/ Akbastau	Kazakhstan	10,737	0.090	Uranium One Inc., KazAtomProm	NA
Budenovskoye 2/Karatau	Kazakhstan	11,232	0.090	KazAtomProm, Uranium One Inc.	NA
Centralnoye (Kanzhagan, S. Muyumkum)	Kazakhstan	25,077	0.070	KazAtomProm	NA
Chiili (North and South Karamurun)	Kazakhstan	27,403	0.070	KazAtomProm	NA
Inkai	Kazakhstan	51,808	0.070	Cameco Corporation, KazAtomProm	2032
Inkai South	Kazakhstan	13,040	0.039	Uranium One Inc., KazAtomProm	NA
Irkol	Kazakhstan	28,641	0.045	KazAtomProm, China Guangdong NPC	NA
Mynkuduk Central	Kazakhstan	48,521	0.040	KazAtomProm	NA

Table 3.   Operating uranium mines and their remaining reasonably assured resources, 2010.—Continued

[tU, metric tonnes; tU/yr, metric tonnes per year; ISL, *in-situ* leach. *In-situ* resources reported, when 2009 Redbook cited information was for production or other descriptive information or where there was no other source of resource data]

Mine/ production center	Production or nominal production capacity/rate (tU/yr)	Geologic type	Type of operation	Information source
Beverley	848	Sandstone	ISL—Acid Leach	Ux Consulting Company LLC, 2010; McKay and Carson, 2010; NEA–IAEA, 2010
Olympic Dam	3,820	Hematite Breccia Complex	Underground— Cu/Au	McKay and Carson, 2010; NEA–IAEA, 2010
Ranger	4,660	Unconformity— Proterozoic	Open Pit	McKay and Carson, 2010; NEA–IAEA, 2010
Lagoa Real-Caetite District	340	Metasomatite	Open Pit	NEA–IAEA, 2010; Ux Consulting Company LLC, 2010
McArthur River	7,200	Unconformity— Proterozoic	Underground	World Nuclear Association, 2011b; Ux Consulting Company LLC, 2010
McClean Lake	3,077	Unconformity— Proterozoic	Underground	World Nuclear Association, 2011b; Ux Consulting Company LLC, 2010
Rabbit Lake/Eagle Point	4,615	Unconformity— Proterozoic	Underground	World Nuclear Association, 2011b; Ux Consulting Company LLC, 2010
Dep. 512/Yining	NA	Sandstone	ISL	NEA–IAEA, 2010
Lantian Deposit	100	Granite or Vein	Underground/ Heap Leach	NEA–IAEA, 2010
Lianshanguan	220	Granite (Benxi) Granite (Qinglong)	Unknown	Dahlkamp, 2010
Qinglong district	100	NA	NA	NEA–IAEA, 2010
Shaoguan	100	Granite	Underground	NEA–IAEA, 2010
Tengchong deposit	NA	Sandstone	ISL	NEA–IAEA, 2010
Xiangshan district	200	Volcanic	Underground/ Mill	NEA–IAEA, 2010; Ux Consulting Company LLC, 2010
Xiazhuang district	NA	Granite or Vein	NA	Ux Consulting Company LLC, 2010
Yili deposit	300	Sandstone or Lignite-type	ISL	Dahlkamp, 2010
Ziyuan	NA	NA	NA	NEA–IAEA, 2010
Rozna	255	Vein	Underground	NEA–IAEA, 2010; Ux Consulting Company LLC, 2010
Straz	38	Sandstone	ISL—Acid Leach	NEA–IAEA, 2010; Ux Consulting Company LLC, 2010
Bagjata	Part of Jaduguda	Vein	Underground	Chaki, 2010
Bhatin	Part of Jaduguda	Vein	Underground	NEA–IAEA 2010; Ux Consulting Company LLC, 2010
Jaduguda	175	Vein	Underground	NEA–IAEA, 2010; Ux Consulting Company LLC, 2010
Narwapahar	Part of Jaduguda	Vein	Underground	NEA–IAEA, 2010; Ux Consulting Company LLC, 2010
Turamdih/Banduhurang	190	Vein	Underground	NEA–IAEA, 2010; Ux Consulting Company LLC, 2010
Gachin (Bandar-Abas)	21	Surficial	Open Pit	NEA–IAEA, 2010
Akdala	1,000	Sandstone	ISL—Acid Leach	NEA–IAEA, 2010; Ux Consulting Company LLC, 2010
Budenovskoye 1-3-4/ Akbastau	3,000	Sandstone	ISL—Acid Leach	NEA–IAEA, 2010; Ux Consulting Company LLC, 2010
Budenovskoye 2/Karatau	1,000	Sandstone	ISL—Acid Leach	NEA–IAEA, 2010; Ux Consulting Company LLC, 2010
Centralnoye (Kanzhagan, S. Muyumkum)	1,000	Sandstone	ISL—Acid Leach	NEA–IAEA, 2010; Ux Consulting Company LLC, 2010
Chiili (North and South Karamurun)	1,000	Sandstone	ISL—Acid Leach	NEA–IAEA, 2010; Ux Consulting Company LLC, 2010
Inkai	2,000	Sandstone	ISL—Acid Leach	NEA–IAEA, 2010; Ux Consulting Company LLC, 2010
Inkai South	2,000	Sandstone	ISL—Acid Leach	NEA–IAEA, 2010; Ux Consulting Company LLC, 2010
Irkol	750	Sandstone	ISL—Acid Leach	NEA–IAEA, 2010; Ux Consulting Company LLC, 2010
Mynkuduk Central	2,000	Sandstone	ISL—Acid Leach	NEA–IAEA, 2010; Ux Consulting Company LLC, 2010

Table 3.   Operating uranium mines and their remaining reasonably assured resources, 2010.—Continued

[tU, metric tonnes; tU/yr, metric tonnes per year; ISL, *in-situ* leach. *In-situ* resources reported, when 2009 Redbook cited information was for production or other descriptive information or where there was no other source of resource data]

Mine/ production center	Country	Remaining *in-situ* reserve tU (proven and probable economic)	Grade %U	Ownership	End of mine life
Mynkuduk West	Kazakhstan	24,540	0.040	KazAtomProm, Sumitomo Corporation, Kansai Electric Power Company	NA
Semisbai	Kazakhstan	17,099	0.070	KazAtomProm, China Guangdong NPC	NA
Stepnoye (Uvanas, East Mynkuduk)	Kazakhstan	22,972	0.036	KazAtomProm	NA
Tortkuduk/Muyunkum (Katco JV)	Kazakhstan	24,131	0.090	Areva NC, KazAtomProm	NA
Tselinny Mining and Chemical Combine	Kazakhstan	9,730	0.160	KazAtomProm	NA
Zarechnoye	Kazakhstan	12,618	0.056	Uranium One Inc., KazAtomProm, Kara-Balta Ore Mining Combine	NA
Kayelekera	Malawi	11,265	0.034	Paladin Energy Ltd.	2020
Langer Heinrich	Namibia	60,830	0.055	Paladin Energy Ltd.	2026
Rossing	Namibia	50,657	0.031	Rio Tinto, Other interests, IDC of S. Africa, Namibian Govt.	2023
Akouta	Niger	24,670	0.34–0.47	Areva NC, Office National des Resources Mineres, Overseas Uranium Resource Development Company Ltd., ENUSA Industrias Avanzadas S.A.	NA
Arlit	Niger	23,171	0.22–.29	Areva NC, Office National des Resources Mineres	NA
Dera Ghazi Khan	Pakistan	NA	0.100	Pakistan Atomic Energy Commission	NA
Issa Khel/Qabul Khel	Pakistan	NA	0.500	Pakistan Atomic Energy Commission	NA
Tumman Leghari	Pakistan	NA	NA	Pakistan Atomic Energy Commission	NA
Crucea, Bihor and Crucea	Romania	8,769	0.210	Uranium National Company (UNC) Romania	2020
Dalur Production Center (Dalmatovskoye deposit)	Russian Federation	10,423	0.041	ARMZ (Atomredmetzoloty OJSC)	NA
Streltsovskoye	Russian Federation	118,341	0.180	ARMZ (Atomredmetzoloty OJSC)	NA
Ezulwini	South Africa	2,730	0.023	First Uranium Corporation	NA
Vaal River Complex	South Africa	14,346	0.016–0.061	Nufcor International Ltd.	NA
Michurinskoye (Inglul'skaya)	Ukraine	25,154	0.100	Vostochny Integrated Mining and Concentrating Plant	NA
Vatutinskoye (Smolinskaya)	Ukraine	23,692	0.100	Vostochny Integrated Mining and Concentrating Plant	NA
Alta Mesa	USA	2,923	0.080	Mestena Uranium LLC	>2017
AZ1	USA	330	0.600	Denison Mines Corp.	NA
Beaver +Pandora	USA	2,038	0.210	Denison Mines Corp.	NA
Crow Butte	USA	1,577	0.130	Cameco Corporation	>2017
Daneros	USA	462	0.280	White Canyon Uranium Ltd.	2014
La Palangana	USA	407	0.110	Uranium Energy Corp.	NA
Smith Ranch–Highland (+Reynolds)	USA	2,269	0.100	Cameco Corporation	NA
White Mesa (Mill)	USA	NA	variable	Denison Mines Corp.	NA
All regions (Centres 1,2,3)	Uzbekistan	76,000	0.02– 17	Navoi Mining and Metallurgy Combinat	2040
Total		1,397,822		Known Production Capacity	

Table 3.   Operating uranium mines and their remaining reasonably assured resources, 2010.—Continued

[tU, metric tonnes; tU/yr, metric tonnes per year; ISL, *in-situ* leach. *In-situ* resources reported, when 2009 Redbook cited information was for production or other descriptive information or where there was no other source of resource data]

Mine/ production center	Production or nominal production capacity/rate (tU/yr)	Geologic type	Type of operation	Information source
Mynkuduk West	1,000	Sandstone	ISL—Acid Leach	NEA–IAEA, 2010; Ux Consulting Company LLC, 2010
Semisbai	500	Sandstone	ISL—Acid Leach	NEA–IAEA, 2010; Ux Consulting Company LLC, 2010
Stepnoye (Uvanas, East Mynkuduk)	1,300	Sandstone	ISL—Acid Leach	NEA–IAEA, 2010; Ux Consulting Company LLC, 2010
Tortkuduk/Muyunkum (Katco JV)	4,000	Sandstone	ISL—Acid Leach	NEA–IAEA, 2010; Ux Consulting Company LLC, 2010
Tselinny Mining and Chemical Combine	500	Vein	Underground— Heap Leach	NEA–IAEA, 2010; Ux Consulting Company LLC, 2010
Zarechnoye	2,000	Vein	Underground— Heap Leach	NEA–IAEA, 2010; Ux Consulting Company LLC, 2010
Kayelekera	1,270	Sandstone	Open Pit	NEA–IAEA, 2010; Ux Consulting Company LLC, 2010
Langer Heinrich	1,425	Surficial	Open Pit	NEA–IAEA, 2010; Ux Consulting Company LLC, 2010
Rossing	3,817	Intrusive	Open Pit	Ux Consulting Company LLC, 2010
Akouta	2,000	Sandstone	Open Pit	NEA–IAEA, 2010; Ux Consulting Company LLC, 2010
Arlit	2,000	Sandstone	Open Pit	NEA–IAEA, 2010; Ux Consulting Company LLC, 2010
Dera Ghazi Khan	30	Sandstone	Unknown	NEA–IAEA, 2010
Issa Khel/Qabul Khel	NA (Heap Leach)	Sandstone	Open Pit/ Heap Leach	IAEA, 2009, Ux Consulting Company LLC, 2010
Tumman Leghari	NA	NA	Unknown	IAEA, 2009, Ux Consulting Company LLC, 2010
Crucea, Bihor and Crucea	300	Vein	Underground	NEA–IAEA , 2010; Ux Consulting Company LLC, 2010
Dalur Production Center (Dalmatovskoye deposit)	3,000	Sandstone	ISL	NEA–IAEA, 2010
Streltsovskoye	3,000	Volcanic	Underground	NEA–IAEA, 2010; Ux Consulting Company LLC, 2010
Ezulwini	462	Quartz-pebble Conglomerate	Underground	NECSA, 2010; Ux Consulting Company LLC, 2010
Vaal River Complex	3,400	Quartz-pebble Conglomerate	Underground	NECSA, 2010; Ux Consulting Company LLC, 2010
Michurinskoye (Inglul'skaya)	1,500	Metasomatite	Underground	NEA–IAEA, 2010
Vatutinskoye (Smolinskaya)	NA	Metasomatite	Underground	NEA–IAEA, 2010
Alta Mesa	385	Sandstone	ISL— Alkaline Leach	Energy Information Administration, 2010a; Ux Consulting Company LLC, 2010
AZ1	White Mesa Mill	Collapse Breccia Pipe	Underground	Pool and Ross, 2007
Beaver +Pandora	White Mesa Mill	Sandstone	Underground	Ux Consulting Company LLC, 2010
Crow Butte	385	Sandstone	ISL— Alkaline Leach	Cameco Corporation, 2011
Daneros	White Mesa Mill	Sandstone	Underground	Ux Consulting Company LLC, 2010
La Palangana	385	Sandstone	ISL— Alkaline Leach	Energy Information Administration, 2010a; Rigby, 2010
Smith Ranch - Highland (+Reynolds)	2,116	Sandstone	ISL— Alkaline Leach	Cameco, 2010
White Mesa (Mill)	1,200	Various sources	Mill	EIA, 2010
All regions (Centres 1,2,3)	NA	Sandstone	ISL	NEA–IAEA, 2005, 2010
Total	75,984			

*No available information, used estimates from NEA–IAEA (2010) which probably do not account for depletion by mining.

#Updated grades not available, used information from IAEA, 2010.

Table 4. Uranium properties that are permitted, or in feasibility or in the prefeasibility stages and their reported resources, 2010

[UG, underground; ISL, *in-situ* leach; OP, open pit; R, resource only, not reserve; I, indicated resource; NA, data not available; tU/yr, metric tonnes per year. Values used because the unique character of the deposits and proposed mining method will likely not yield a reserve prior to development. *In-situ* resources reported, when 2009 Redbook cited information was for production or other descriptive information or where there was no other source of resource data]

Mine name	Country	In-situ resource	Grade	Status	Ownership
Cerro Solo	Argentina	3,900	0.4	Feasibility study	Commision Nacional de Energia Atomica
Sierra Pintada	Argentina	2,620	0.1	Feasibility study	Commision Nacional de Energia Atomica
Four Mile	Australia	23,462	0.28	Feasibility study	Quasar Resources Pty. Ltd.
Honeymoon	Australia	2,500	0.2035	Development	Uranium One Inc.
Lake Maitland	Australia	10,000	0.031	Development	Mega Uranium Ltd.
Oban	Australia	1,781	NA	Feasibility study	Curnamona Energy Ltd
Wiluna	Australia	9,385	0.021–.050	Feasibility study	Toro Energy Ltd.
Yeleerie	Australia	44,077	0.15	Feasibility study	BHP Billiton
Itataia-Santa Quiteria District	Brazil	67,240	0.08	Feasibility study	Industrias Nucleares do Brazil
Cigar Lake	Canada	80,500	16.59	Development	Cameco Corporation
Kiggavik	Canada	51,574	0.22	Feasibility study	Areva Resources Canada Inc.
Matoush	Canada	7,770	0.45	Feasibility study	Strateco Resources Inc.
Michelin & Jacques Lake	Canada	29,923	0.49	Feasibility study	Aurora Energy Resources Ltd.
Midwest	Canada	16,340	4.4	Feasibility study	Areva Resources Canada Inc.
Millenium	Canada	18,002	3.8	Feasibility study	Cameco Corporation
Bakouma	Central African Republic	9,885	1.72	Feasibility study	Areva Resources Centrafrique
Dongsheng	China	5,000	NA	Feasibility study	China National Nuclear Association
Erdos	China	21,600	NA	Feasibility study	China National Nuclear Association
Erlian	China	19,400	Unknown	Feasibility study	China National Nuclear Association
Guyuan	China	5,000	0.1–0.3	Feasibility study	China National Nuclear Association
Liaohe	China	NA	NA	Feasibility study	China National Nuclear Association
Shihongtan deposit	China	3,000	0.03	Feasibility study	China National Nuclear Corporation
Turp-Hame	China	9,000	NA	Feasibility study	China National Nuclear Association
Zaohuohao	China	17,000	Unknown	Feasibility study	China National Nuclear Association
Talvivaara	Finland	17,110	0.0018	Development	Talvivaara Mining Company Plc.
Tummalapalle - Rachakuntapalle	India	12,555	0.04	Committed	Uranium Corporation of India
Mohuldih	India	Unknown	Unknown	Committed	Uranium Corporation of India
Lambapur-Peddagattu	India	Unknown	Unknown	Committed	Uranium Corporation of India
Saghand (Ardakan)	Iran	900	0.055	Development	Atomic Energy Organization of Iran
Kharasan 1 North	Kazakhstan	34,350	0.11	Development	Energy Asia Ltd., Kazatomprom, Uranium One Inc.
Kharasan-2 South	Kazakhstan	24,751	0.1–.2	Development	Energy Asia Ltd, KazAtomProm
Zhalpak	Kazakhstan	15,000	NA	Committed	Zhalpak JV, KazAtomProm
Dornod District (12 deposits)	Mongolia	24,780	0.116	Feasibility study	Khan Resources Inc, Priargunsky Chemical and Mining enterprise, Mongol Erdene Holding Co.
Marenica	Namibia	62,856	0.017	Pre-feasability	Marenica Energy Ltd.
Trekkopje/ Klein Trekkopje	Namibia	43,243	0.015	Development	Areva Resources Namibia
Valencia	Namibia	23,269	0.019	Development	Forsys Metals Corp.
Azelik	Niger	10,800	NA	Development	China U International Uranium Corporation, ZX Joy Global Inc., Société de Patrimoine des Mines du Niger, Trendfield Holdings SA
Imouraren	Niger	183,520	.046–0.11	Development	Areva Resources Namibia, Société de Patrimoine des Mines du Niger , Kansai Electric Power Co. Inc.
Shanawah	Pakistan	2,578	0.05	Development	Pakistan Atomic Energy Commission
Elkon (Yuzhnoe, Severnoe)	Russian Federation	71,300	0.15	Development	ARMZ (Atomredmetzoloty OJSC)
Gornoe, Beryozovoe	Russian Federation	7,918	0.2	Development	ARMZ (Atomredmetzoloty OJSC)
Khiagda, Vershinnoe	Russian Federation	26,805	0.055	Development	ARMZ (Atomredmetzoloty OJSC)
Olovskoye	Russian Federation	12,200	0.082	Development	ARMZ (Atomredmetzoloty OJSC)

Table 4. Uranium properties that are permitted, or in feasibility or in the prefeasibility stages and their reported resources, 2010.—Continued

[UG, underground; ISL, *in-situ* leach; OP, open pit; R, resource only, not reserve; I, indicated resource; NA, data not available; tU/yr, metric tonnes per year. Values used because the unique character of the deposits and proposed mining method will likely not yield a reserve prior to development. *In-situ* resources reported, when 2009 Redbook cited information was for production or other descriptive information or where there was no other source of resource data]

Mine name	Geologic type	Projected start-up	Proposed mining method	Proposed capacity (tU/yr)	Information source
Cerro Solo	Sandstone	NA	OP	NA	IAEA, 2010; NEA–IAEA , 2010; Ux Consulting Company LLC, 2010
Sierra Pintada	Volcanic	NA	OP	NA	IAEA, 2010; NEA–IAEA , 2010; Ux Consulting Company LLC, 2010
Four Mile	Sandstone	2011	ISL	1,000	NEA–IAEA , 2010; Ux Consulting Company LLC, 2010
Honeymoon	Sandstone	2010	ISL	340	NEA–IAEA , 2010; Ux Consulting Company LLC, 2010
Lake Maitland	Surficial	NA	OP	NA	Mega Uranium, 2011; Ux Consulting Company LLC, 2010
Oban	Sandstone	NA	ISL	NA	NEA–IAEA , 2010; Ux Consulting Company LLC, 2010
Wiluna	Surficial	NA	OP	NA	Toro Energy Ltd., 2011; Ux Consulting Company LLC, 2010
Yeleerie	Surficial	NA	OP	NA	BHPBilliton, 2011; Ux Consulting Company LLC, 2010
Itataia-Santa Quiteria District	Phosphorite	2012	OP	1,000	NEA–IAEA , 2010; Ux Consulting Company LLC, 2010
Cigar Lake	Unconformity-Proterozoic	2012	UG	6,249	NEA–IAEA , 2010; Ux Consulting Company LLC, 2010
Kiggavik	Unconformity-Proterozoic	NA	UG	NA	World Nuclear Association, 2011; Ux Consulting Company LLC, 2010
Matoush	Vein	NA	UG	NA	Calvert, 2010; Ux Consulting Company LLC, 2010
Michelin & Jacques Lake	Volcanic	NA	OP/UG	NA	Calvert, 2010; Ux Consulting Company LLC, 2010
Midwest	Unconformity-Proterozoic	2013	OP	2,300	World Nuclear Association, 2011b; Ux Consulting Company LLC, 2010; NEA–IAEA, 2010
Millenium	Unconformity-Proterozoic	NA	UG	5,700	Calvert, 2010; Ux Consulting Company LLC, 2010
Bakouma	Phosphoritic sandstone	2015	OP	1,200	Ux Consulting Company LLC, 2010; Areva, 2011
Dongsheng	Sandstone	NA	ISL/OP	NA	NEA–IAEA , 2010
Erdos	Sandstone	NA	NA	NA	Ux Consulting Company LLC, 2010
Erlian	Sandstone	NA	UG	NA	Zhang, 2010; NEA–IAEA , 2010
Guyuan	Granite	NA	ISL	NA	Dahlkamp, 2010
Liaohe	Sandstone	NA	NA	NA	Zhang, 2010; NEA–IAEA , 2010
Shihongtan deposit	Sandstone	NA	ISL	NA	NEA–IAEA , 2008
Turp-Hame	Sandstone	NA	NA	NA	NEA–IAEA , 2010
Zaohuohao	Unknown	NA	ISL	NA	NEA–IAEA , 2008
Talvivaara	Black Shales	2012	By-product	350	Talvivaara Mining Company Plc., 2011
Tummalapalle - Rachakuntapalle	Strata-bound	2010	UG	217	Chaki, 2010
Mohuldih	Vein	2011	UG	190	Chaki, 2010
Lambapur-Peddagattu	NA	2012	UG-OP	130	NEA–IAEA , 2010; Ux Consulting Company LLC, 2010
Saghand (Ardakan)	Metasomatite	2012	UG	50	NEA–IAEA , 2010
Kharasan 1 North	Sandstone	2012	ISL	2,000	NEA–IAEA , 2010; Ux Consulting Company LLC, 2010
Kharasan-2 South	Sandstone	2012	ISL	2,000	NEA–IAEA , 2010; Ux Consulting Company LLC, 2010
Zhalpak	Sandstone	2015	ISL	1,000	NEA–IAEA , 2010
Dornod District (12 deposits)	Volcanic	2015	OP,UG	1,150	NEA–IAEA , 2010; Ux Consulting Company LLC, 2010
Marenica	Surficial	2013	OP	1,000	Marencia Energy Ltd, 2011; NEA–IAEA , 2010; Ux Consulting Company LLC, 2010
Trekkopje/ Klein Trekkopje	Surficial	2016	OP	1,600	NEA–IAEA , 2010; Ux Consulting Company LLC, 2010
Valencia	Surficial	2013	OP	1,400	NEA–IAEA , 2010; Ux Consulting Company LLC, 2010
Azelik	Sandstone	2011	OP	1,000	NEA–IAEA , 2010; Ux Consulting Company LLC, 2010
Imouraren	Sandstone	2012	OP	5,000	NEA–IAEA , 2010; Ux Consulting Company LLC, 2010
Shanawah	Sandstone	2014	ISL	50	IAEA, 2010; NEA–IAEA, 2010
Elkon (Yuzhnoe, Severnoe)	Metasomatite	2016	UG	5,000	Ux Consulting Company LLC, 2010
Gornoe, Beryozovoe	Vein	2014	UG	600	NEA–IAEA, 2010
Khiagda, Vershinnoe	Sandstone	2011	ISL	1,000	NEA–IAEA, 2010
Olovskoye	Vein	2014	OP,UG	600	NEA–IAEA, 2010; Ux Consulting Company LLC, 2010

Table 4.   Uranium properties that are permitted, or in feasibility or in the prefeasibility stages and their reported resources, 2010.—Continued

[UG, underground; ISL, *in-situ* leach; OP, open pit; R, resource only, not reserve; I, indicated resource; NA, data not available; tU/yr, metric tonnes per year. Values used because the unique character of the deposits and proposed mining method will likely not yield a reserve prior to development. *In-situ* resources reported, when 2009 Redbook cited information was for production or other descriptive information or where there was no other source of resource data]

Mine name	Country	In-situ resource	Grade	Status	Ownership
Beatrix	South Africa	24,600	NA	Development	Harmony Gold Mining Co.
Cooke Dump	South Africa	9,464	0.09	Feasibility study	Harmony Gold Mining Co.
Dominion Reef	South Africa	55,753	0.062	Development	Shiva Uranium Pty. Ltd.
Henkries	South Africa	1,145	NA	Development	Niger Uranium Ltd.
Klerksdorp and Southern Free	South Africa	2,972	0.02	Development	Witwatersrand Consolidated Gold Resources Ltd.
Ezulwini (Randfontein)	South Africa	2,692	0.007	Development	First Uranium Corp.
Ryst Kuil	South Africa	7,731	0.1	Development	Gold Fields Ltd., Areva Resources Southern Africa
Western Rand tailings	South Africa	11,387	0.004	Development	Mintails Ltd.
MWS Tailings	South Africa	9,269	0.001	Development	First Uranium Corp.
Salamanca I	Spain	30,926	0.0563	Feasibility study	Berkeley Resources, ENUSA Industrias Avanzadas S A.
Novokonstantinovskoe	Ukraine	93,630	0.139	Development	Vostochny Integrated Mining and Concentrating Plant
Safonovskoye	Ukraine	6,900	0.035	Development	Vostochny Integrated Mining and Concentrating Plant
Bullfrog	USA	1,798	0.33	Development	Denison Mines Corp.
Canyon	USA	586	0.92	Development	Denison Mines Corp.
Centennial	USA	3,989	0.08	Development	Powertech Uranium Corp.
Willow Creek	USA	7,506	0.09	Operational	Uranium One Inc.
Church Rock	USA	7,154	0.1	Partially permitted and licensed	Hydro Resources Inc.
Crownpoint	USA	5,885	0.16	Partially permitted and licensed	Hydro Resources Inc.
Dewey Burdock	USA	2,571	0.15	Development	Powertech Uranium Corp.
Gas Hills-peach	USA	5,270	0.13	Development	Cameco Corporation
Goliad	USA	2,106	0.42	Partially permitted and licensed	Uranium Energy Corp.
Hank	USA	860	0.1	Development	Uranerz Energy Corp
Lance	USA	3,539	0.039	Feasibility study	Peninsula Energy
Lost Creek	USA	3,769	0.044	Development	Ur Energy Inc.
Moore Ranch	USA	2,230	0.085	Permitted and licensed	Uranium One Inc.
Nichols Ranch	USA	1,115	0.1	Partially permitted	Uranerz Energy Corp.
North Butte-Brown Ranch	USA	3,154	0.1	Development	Cameco Corporation
Pinenut (AZ Strip)	USA	336	0.4	Development	Denison Mines Corp.
Reno Creek	USA	1,651	0.05	Development	Uranerz Energy Corp.
Roca Honda	USA	6,730	0.196	Feasibility study	Strathmore Minerals Corp.
Tony M + Southwest	USA	3,131	0.2	Development	Denison Mines Corp.
Whirlwind, Energy Queen, San Rafael	USA	1,334	0.2–0.3	Development	Energy Fuels Inc.
Pinon Ridge Mill	USA	NA	NA	Permitting	Energy Fuels Inc.
Chimiwungo	Zambia	1,020	0.047	Development	Equinox Minerals Ltd.
Total *in-situ* resource (tU)		1,355,097			

Table 4. Uranium properties that are permitted, or in feasibility or in the prefeasibility stages and their reported resources, 2010.—Continued

[UG, underground; ISL, *in-situ* leach; OP, open pit; R, resource only, not reserve; I, indicated resource; NA, data not available; tU/yr, metric tonnes per year. Values used because the unique character of the deposits and proposed mining method will likely not yield a reserve prior to development. *In-situ* resources reported, when 2009 Redbook cited information was for production or other descriptive information or where there was no other source of resource data]

Mine name	Geologic type	Projected start-up	Proposed mining method	Proposed capacity (tU/yr)	Information source
Beatrix	Quartz-pebble conglomerate	NA	UG	NA	NEA–IAEA, 2010
Cooke Dump	Surface dams, dumps and slimes	2012	Surface	NA	NEA–IAEA, 2010
Dominion Reef	Quartz-pebble conglimerate	2012	UG	1,460	NECSA, 2010
Henkries	Surficial	NA	OP	NA	NEA–IAEA, 2010
Klerksdorp and Southern Free	Quartz-pebble conglomerate	NA	UG	NA	NEA–IAEA, 2010; NECSA, 2010
Ezulwini (Randfontein)	Quartz-pebble conglomerate	2012	UG	425	NECSA, 2010
Ryst Kuil	Sandstone	NA	OP	1,136	NECSA, 2010; NEA–IAEA, 2010; Ux Consulting Company LLC, 2010
Western Rand tailings	Quartz-pebble conglomerate	NA	OP	NA	NECSA, 2010
MWS Tailings	Surface tailings	2012	OP	515	UxConsulitng, 2010
Salamanca I	Metasomatite	2014	OP	769	NEA–IAEA, 2010; Ux Consulting Company LLC, 2010
Novokonstantinovskoe	Metasomatite	2011	UG	2,500	NEA–IAEA, 2010; Ux Consulting Company LLC, 2010
Safonovskoye	Sandstone	2012	ISL	210	NEA–IAEA, 2010; NECSA, 2010
Bullfrog	Sandstone	NA	UG	White Mesa Mill	Ux Consulting Company LLC, 2010
Canyon	Breccia Pipe	NA	UG	White Mesa Mill	Ux Consulting Company LLC, 2010
Centennial	Sandstone	2012	OP	269	Ux Consulting Company LLC, 2010
Willow Creek	Sandstone	2011	ISL	385	Energy Information Administration, 2010a; Ux Consulting Company LLC, 2010
Church Rock	Sandstone	NA	ISL	385	Energy Information Administration, 2010a; Ux Consulting Company LLC, 2010
Crownpoint	Sandstone	NA	ISL	385	Ux Consulting Company LLC, 2010
Dewey Burdock	Sandstone	2013	ISL	346	Ux Consulting Company LLC, 2010
Gas Hills-peach	Sandstone	NA	ISL	NA	Ux Consulting Company LLC, 2010
Goliad	Sandstone	NA	ISL	385	Energy Information Administration, 2010a; Ux Consulting Company LLC, 2010
Hank	Sandstone	NA	ISL	NA	Ux Consulting Company LLC, 2010
Lance	Sandstone	2012	ISL	577	Ux Consulting Company LLC, 2010
Lost Creek	Sandstone	NA	ISL	770	Energy Information Administration, 2010a; Ux Consulting Company LLC, 2010
Moore Ranch	Sandstone	2012	ISL	192	Energy Information Administration, 2010a; Ux Consulting Company LLC, 2010
Nichols Ranch	Sandstone	2011	ISL	NA	Ux Consulting Company LLC, 2010
North Butte-Brown Ranch	Sandstone	NA	ISL	NA	Ux Consulting Company LLC, 2010
Pinenut (AZ Strip)	Breccia Pipe	NA	UG	White Mesa Mill	Ux Consulting Company LLC, 2010
Reno Creek	Sandstone	NA	ISL	NA	Ux Consulting Company LLC, 2010
Roca Honda	Sandstone	NA	ISL	NA	Ux Consulting Company LLC, 2010
Tony M + Southwest	Sandstone	Standby	UG	White Mesa Mill	Ux Consulting Company LLC, 2010
Whirlwind, Energy Queen, San Rafael	Sandstone	NA	UG	White Mesa Mill	Ux Consulting Company LLC, 2010
Pinon Ridge Mill	Various	2013	UG	96	Energy Information Administration, 2010a
Chimiwungo	Metamorphic	NA	OP	NA	Titley, 2009
Total *in-situ* resource (tU)	Proposed known production capacity			52,741	

[1]Uses measured and indicated resource (demonstrated economic) estimates when those are available. No cost cutoffs or ranges are included since mines are in planning stages. Estimates for some mines may include inferred resources.

*No available information, used IAEA, 2010 estimates which probably do not account for depletion by mining.

#Updated grades not available, used original IAEA, 2010 values.

Geologic Assurance for and Economic Feasibility of Worldwide Uranium Resources, 2010

Decreasing degree of geological assurance →

		Identified resources (tU)		Undiscovered resources (tU)	
		Demonstrated	Inferred	Prognosticated	Speculative
Economic	RAR < USD 80/KgU	2,516,100	1,225,800	1,701,500	None reported
		(2,100,000 recoverable) 2,700,000 *in situ* In operating or developing mines (Calculated as of 2010 in this report)			
	RAR USD 80–130/KgU	1,008,800	653,300	1,113,300	3,738,200
Subeconomic	RAR USD 130–260/KgU	479,600	422,700	90,200	163,500
					3,593,800 No cost range assigned

Unconventional resources	Uraniferous Phosphate: Demonstrated and Inferred Resource 7.3–7.6 million tU (NEA–IAEA, 2010)
	Black Shale Deposits: 1,061,958 tU—IAEA UDEPO database (IAEA, 2010)
	Lignite Deposits: 224,360 tU—IAEA UDEPO database (IAEA, 2010)
	Seawater: Approximately 4,000,000,000 tU (NEA–IAEA, 2010) (uneconomic)

Decreasing degree of economic feasibility

Secondary Sources:

Enrichment tails	1,600,000 tU	worldwide (production depends on enrichment capacity)
Stockpiles	up to 575,000 tU	(NEA–IAEA estimate)
LEU	11,905 tU/year	(from the HEU–LEU program, expires in 2013; USEC, 2011)

Note: RAR tU is reported as recoverable uranium (milling and mining losses deducted) except for unconventional and secondary sources that are too speculative to treat as RAR. Data from the 2009 Red Book (NEA–IAEA, 2010) except where indicated.

Figure 13. Uranium Resources Mapped to Resource Category Geological assurance for, and economic feasibility of, worldwide uranium resources, by resource categories used by the Organisation for Economic Co-operation Nuclear Agency and International Atomic Energy Agency.

Table 5.   Reasonably assured resources reported in the 2009 Red Book (NEA–IAEA, 2010).

[tU, metric tonnes; kg, kilograms; USD, U.S. dollars]

| | Identified reasonable assured and inferred resources (tU) from the 2009 Red Book (NEA–IAEA, 2010) (rounded to the nearest 100 tU) | | | |
| | Cost ranges | | | |
Country	<USD 40/kg U tU	<USD 80/kg U tU	<USD 130/kg U tU	<USD 260/kg U tU
Algeria[a,b,c]	0*	0*	19,500	19,500
Argentina	0	11,400	19,100	19,100
Australia	NA	1,612,000	1,673,000	1,679,000
Brazil	139,900	231,300	278,700	278,700
Canada	366,700	447,400	485,300	544,700
Central African Republic[a,b,c]	0*	0*	12,000	12,000
Chile[c]	0	0	0*	1,500
China[c]	67,400	150,000	171,400	171,400
Congo, Dem. Rep. of[a,b,c]	0	0*	0*	2,700
Czech Republic	0	500	500	500
Denmark[b,c]	0	0	0	85,600*
Egypt	0	0	0	1,900
Finland[b,c]	0	0	1,100	1,100
France	0	0	100	9,100
Gabon[a,b]	0	0	4,800	5,800
Germany[b,c]	0	0	0	7,000
Greece[a,b]	0*	0*	0*	7,000
Hungary	0	0	0	8,600
India[c,d]	0	0	80,200	80,200
Indonesia[b,c]	0*	0*	4,800	6,000
Iran, Islamic Republic of	0	0	0*	2,200
Italy[a,b]	0	0	4,800	6,100
Japan[b]	0	0*	6,600	6,600
Jordan[a,c]	0*	111,800	111,800	111,800
Kazakhstan[c]	44,400	475,500	651,800	832,000
Malawi*	0	8,100	15,000	15,000
Mexico[a,b,c]	0	0	0*	1,800
Mongolia[b,c]	0	41,800	49,300	49,300
Namibia[a,c]	0*	2,000*	284,200*	284,200*
Niger[a,c]	17,000*	73,400*	272,900*	275,500*
Peru[c]	0	0	2,700	2,700
Portugal[a,b]	0	4,500	7,000	7,000
Romania[a]	0	0	6,700	6,700
Russian Federation	0	158,100	480,300	566,300
Slovakia*	0	0	0	10,200
Slovenia[a,b,c]	0	0*	9,200	9,200
Somalia[a,b,c]	0	0*	0*	7,600
South Africa[b,f]	155,300	232,900	295,600	295,600
Spain[b]	0	2,500	11,300	11,300
Sweden[a,b]	0	0	10,000	10,000
Tanzania[c]	0	0	0	28,400*
Turkey[b,c]	0	0*	7,300	7,300
Ukraine[c]	5,700	53,500	105,000	223,600
United States	0	39,000	207,400	472,100
Uzbekistan[a,c,e]	0	86,200*	114,600*	114,600*
Vietnam[a,b,c]	0	0*	0*	6,400
Zimbabwe[a,b,c]	0	0*	0*	1,400
Total[g]	796,400	3,741,900	5,404,000	6,306,300

*IAEA Uranium Group Secretariat estimate.

[a]Not reported in 2009 responses, data from previous Red Book.

[b]Assessment not made within the last five years.

[c]*In-situ* resources were adjusted by the Secretariat to estimate recoverable resources using recovery factors provided by countries or estimated by the Secretariat according to the expected production method.

[d]Cost data are not provided, therefore resources are reported in the <USD 130/kgU category.

[e]Data from 2007 Red Book (NEA–OECD, 2008), reduced by past production.

[f]Resource estimates do not account for production.

[g]Totals related to cost ranges <USD 40/kgU and <USD 80/kgU are higher than reported in the tables because certain countries do not report resource estimates, mainly for reasons of confidentiality.

Table 6.   Reasonably assured resources (RAR) and production capacity of operating and proposed mines by extraction technology.

[tU, metric tonnes; tU/yr, metric tonnes per year; ISL, *in-situ* leaching]

	Reasonably assured *in-situ* resources (tU)	Nominal production capacity (tU/yr)
Operating mines		
ISL mines	443,600	28,000
Conventional mines	954,200	48,000
Developing mines		
ISL mines	221,700	11,000
Conventional mines	1,124,000	42,000

Resources and capacity rounded to the nearest 100 tU. Production capacity for developing mines is estimated.

Largest Advanced-stage Projects and Operating Mines

A number of mines are expected to continue to produce uranium, and several new mines with significant resources are projected to come online within the 5 years 2012–2016 (tables 3, 4). Production rates for future mines are uncertain, and the mines with the largest resources may not have the highest production rate. Expansion of the Olympic Dam mining operation in Australia, pending financing, would potentially be the largest resource and producer, increasing current production sixfold (Mckay and Carson, 2010). In August 2012 BHP Billiton announced they would be delaying the expansion of Olympic Dam indefinitely (ABC News, 2012). The world's second largest resource after Olympic Dam, the Imouraren mine in Niger, is expected to begin producing uranium in 2014, and to continue for 35 years, through 2045. Canada's McArthur River mine, with the world's third largest reserve, has plans to continue production until 2030. The fourth largest deposit, Priargunsky/Streltsovskoye in Russia, is expected to continue as a producer. The Novokonstantinovskoe mine in Ukraine opened in 2011, bringing a large resource to the world market through a significant investment of $500 million from Russia. Although a series of mine floods have delayed development of Cigar Lake mine, also in Canada, production is expected to begin in 2013. Uzbekistan's ISL properties are expected to continue to produce. Uzbekistan's official program has a goal of increasing State production by 50 percent by 2012, but this increase is dependent on financing of State-owned mines. The Russian Elkon deposit is reported to be targeting uranium production by 2016. Brazil is working to bring Itatiaia–Santa Quitéria, a large unconventional uraniferous phosphate resource, to production by 2013. Namibia will continue as a major producer, with continued mining at the Rossing and the Langer Heinrich deposits. The Dominion Reef mine in South Africa is scheduled to begin to produce in the near future following

a change in ownership from Uranium One to Shiva Uranium Pty. Ltd. Kazakhstan's two largest resource projects, Katco and Inkai, are expected to continue to produce; it is anticipated Inkai will be mined until at least 2032. The Areva NC (Areva) Kiggavik mine in Canada is projected to produce by 2017 or 2018, although the production timeline is uncertain because this project is in the feasibility stage. BHP Billiton appeared to be moving forward in developing the surficial Yeelirrie mine after a mining ban in Western Australia was lifted in 2008, making exploitation possible. However, in 2010 BHP Billiton announced that it would be putting the environmental permitting process on hold, delaying production, and sold the property to Cameco in 2012. Areva's Trekkopje deposit in Niger is currently suspended pending increased uranium prices after the resource estimate was reduced because chemical assays did not confirm the uranium content projected by radiometric surveys (Areva NC, 2011).

Undiscovered Resources

Undiscovered resources both prognosticated and speculative, are those that, on the basis of earlier discoveries in similar geologic settings, are expected to be found. "Prognosticated resources" have some direct evidence for their occurrence, and "speculative resources" are expected to occur, but with no direct evidence that they exist. Thirty three countries have historically reported some undiscovered resources to NEA (table 7). Potentially economic prognosticated resources (extractable for <USD 130/kgU) total 2.8 million tU, and economic speculative resources total 3.7 million tU (NEA–IAEA, 2010). Subeconomic undiscovered resources (prognosticated and speculative), including those (3.5 million tU) for which no cost range has been assigned, total 7.5 million tU.

The global distribution of reported undiscovered resources in many cases reflects the intensity and scope of past and (or) current government or industry mineral exploration programs. The United States reports the largest prognosticated resource (1.2 million tU), followed by Kazakhstan (500,000 tU) and Brazil (300,000 tU) (fig. 14). Mongolia and the United States both report 1.3 million tU in the speculative resource category, South Africa 1.1 million tU, Canada 700,000 tU and Brazil 500,000 tU (fig. 15). Other countries report smaller undiscovered resources. The U.S. National Uranium Resource Evaluation (NURE) program, one of the most active programs for evaluating undiscovered resources, may in part account for its large number of undiscovered resources; the program ended in 1982. Conversely, resource-rich Australia and Namibia do not calculate undiscovered resources at a national scale, and so their potential resources are unreported.

Table 7. Undiscovered resources, in thousands of tonnes uranium metal, as reported in the "2009 Red Book" (NEA–IAEA, 2010).

[NA, data not available; kgU, kilograms uranium metal]

Country	Prognosticated resources Cost ranges		Speculative resources Cost ranges		
	<USD 80/kgU	<USD 130/kgU	<USD 130/kgU	Cost range unassigned	Total
Argentina	1.4	1.4	NA	NA	NA
Brazil	300.0	300.0	NA	500.0	500.0
Canada	50.0	150.0	700.0	0.0	700.0
Chile	NA	1.5	NA	3.2	3.2
China	3.6	3.6	4.1	0.0	4.1
Colombia	NA	NA	NA	20.0	20.0
Czech Republic	0.2	0.2	0.0	179.0	179.0
Denmark[a]	0.0	0.0	50.0	10.0	60.0
Germany	0.0	0.0	0.0	74.0	74.0
Greece[a]	6.0	6.0	0.0	0.0	0.0
Hungary	0.0	18.4	NA	NA	NA
India	NA	50.9	NA	17.0	17.0
Indonesia[a]	NA	NA	0.0	12.5	12.5
Iran, Islamic Republic of	0.0	4.1	12.2	NA	12.2
Italy[a]	NA	NA	NA	10.0	10.0
Jordan	67.8	84.8	84.8	NA	84.8
Kazakhstan	280.0	300.0	500.0	NA	500.0
Mexico[a]	NA	3.0	NA	10.0	10.0
Mongolia[a]	0.0	0.0	1,390.0	NA	1,390.0
Niger[a]	14.5	24.6	NA	NA	NA
Peru	6.6	6.6	19.7	0.0	19.7
Portugal	1.0	1.5	NA	0.0	NA
Romania[a]	NA	3.0	3.0	0.0	3.0
Russian Federation	276.5	276.5	714.0	0.0	714.0
Slovenia	0.0	1.1	NA	NA	NA
South Africa	34.9	110.3	NA	1,112.9	1,112.9
Ukraine	4.0	11.3	120.0	135.0	255.0
United States[b]	839.0	1,273.0	858.0	482.0	1,340.0
Uzbekistan[a]	56.3	85.0	0.0	134.7	134.7
Venezuela[a]	NA	NA	0.0	163.0	163.0
Vietnam	0.0	7.9	100.0	130.0	230.0
Zambia[a]	0.0	22.0	NA	NA	NA
Zimbabwe[a]	0.0	0.0	25.0	0.0	25.0
Total (reported by countries)**	**1,941.8**	**2,746.6**	**4,580.8**	**2,993.3**	**7,574.1**

*Undiscovered resources are reported as *in-situ* resources.

**Totals may not equal sum of components due to independent rounding.

[a]Not reported in 2009 reponses, data from previous Red Book.

[b]The United States does not report inferred or prognosticated resources.

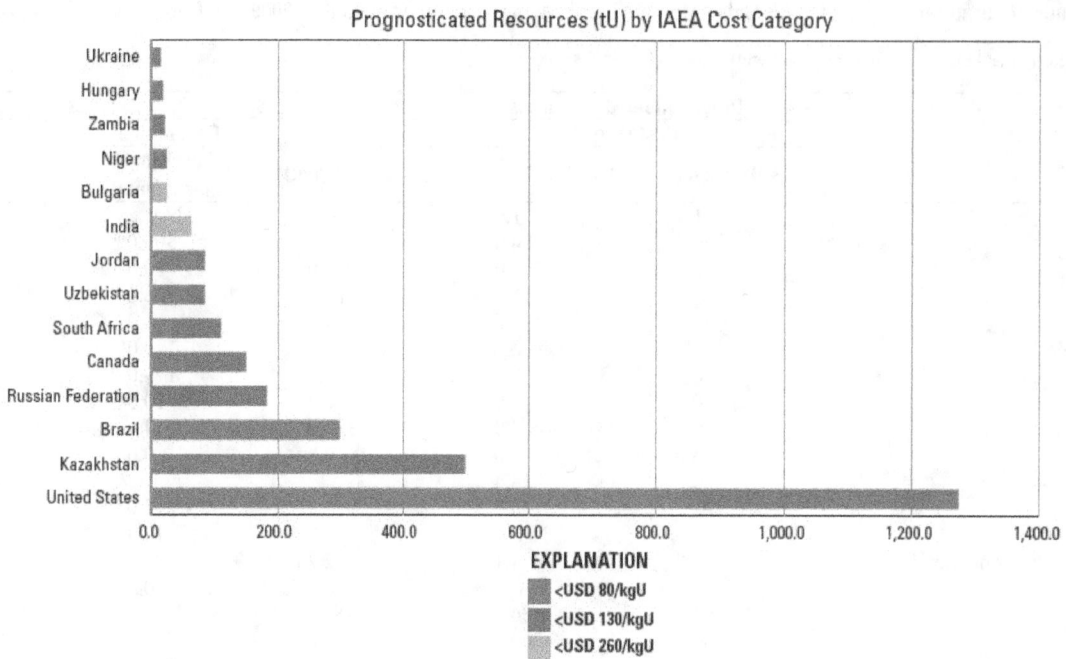

Figure 14.   Prognosticated resources of countries reported as having more than 11,000 tU. From NEA–IAEA (2010).

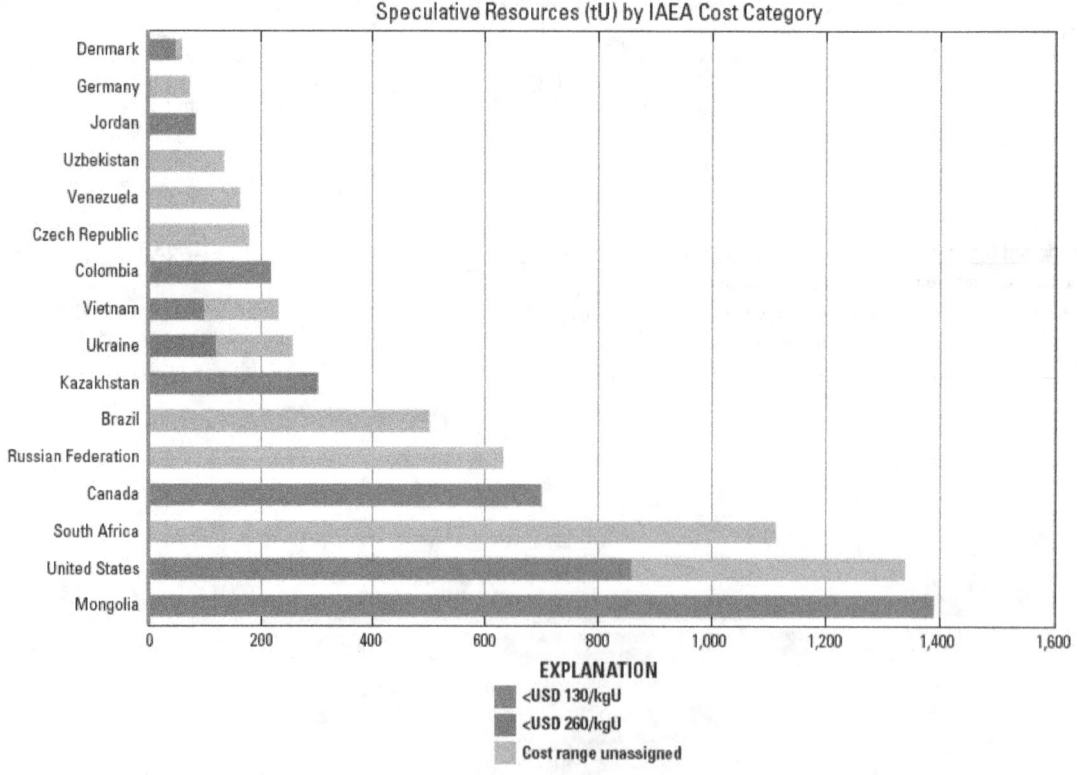

Figure 15.   Speculative resources for countries reporting more than 60,000 tU. Data from NEA–IAEA (2010).

Resources classified as undiscovered should be considered conjectural because the methodology and uncertainties of many undiscovered resource estimating techniques are poorly understood. For this reason, reliance upon undiscovered resources in developing energy policy should be done cautiously. Intensive exploration efforts will be required to move these resources into RAR categories in order to conduct prefeasibility and feasibility studies that could lead to mining.

Historically, the infusion of expenditures for exploring resources yielded a fairly predictable increase in the identification of RAR. Past periods of intense exploration, such as during the 1940s and 1970s, caused by increased demand, led to defining resources in excess of market requirements. Although increased expenditures for exploration uncovered new resources historically, such investment in the 21st century may not yield similar finds. Many of the more easily identified deposits have been discovered and developed, whereas potential deposits are expected to be deeper and to require more innovative exploration techniques to delineate them, as well as more time, expense, and drilling.

Production Capacity

The known nominal or licensed production capacity of currently (during 2010) producing properties is 75,984 tU/yr, slightly greater than the current consumption of 68,971 tU/yr. However, production in 2010 at 53,633 tU, was 70 percent of capacity. Disruptions in large mining operations are not uncommon, and the NEA reports that world production has never exceeded 89 percent of production capacity (Nuclear Energy Agency, 2006). Because mines never produce at or above production capacity for their entire mine life, when predicting future capacity, licensed capacity should be considered a maximum potential capacity.

Proposed mining capacity for developing mines is not always reported. Twenty six of the 74 developing mines reported in table 4 do not have an associated proposed production capacity that has been reported. Analysis of the proposed capacity for those mines for which information is available shows that a yearly minimum production rate of 52,741 tU has a good chance of being developed within the next 5 years.

Uncertainties in Current and Future Production

Mine Floods and Accidents

Current world uranium supply from primary sources is dominated by a few large mines. If there are technical difficulties at any of these large mines, the impact on uranium supply could be profound. At the McArthur River mine, which accounted for 15 percent of world production in 2009, a flood

in 2003 halted production for 4 months delaying production of 1,300 tU. If flooding had not been quickly brought under control, the mine would have been closed for at least a year, according to Ux Consulting Company LLC (2010). The nearby Cigar Lake mine was flooded in 2006, delaying startup until at least 2013, and delaying production of an estimated 7000 tU/yr. Production at Olympic Dam, the fifth largest uranium producer worldwide in 2009, was interrupted that same year by an ore haulage accident, which forced the mine to work on a limited capacity from October 2009 to July 2010, and reducing production by an estimated 1,600 tU (Ux Consulting Company LLC, 2010).

The Influence of Infrastructure on Mine Development and on Definition of Resources

Mine life for the producing mines is difficult to predict. Some operating mines, such as Rabbit Lake in Canada, may soon close if new RAR are not identified. At the other end of the spectrum, Olympic Dam was planning a major expansion that would have extended mine life to 2032, until it announced deferment of these plans in August, 2012.

The availability of local infrastructure can have a strong impact on identifying new RAR at an operating mine site. If existing infrastructure is adequate to support an operating mine, RAR are likely to grow and the life of the mine will be extended. At the Rabbit Lake mine in northern Saskatchewan, Canada, considerable infrastructure was developed to exploit the original 15,770 tU identified at this deposit in 1968. Incremental exploration followed by expansion along the mineralized trend have quadrupled the original capacity of this resource, expanding it to 68,467 tU and extending mine life 27 years beyond the original projected closure.

If infrastructure is lacking, identified deposits will go unmined. Only one uranium mill is currently operating in the western United States, the White Mesa Mill in southeastern Utah, owned and operated by Energy Fuels. Active mines in the Uravan mineral belt of Colorado and Utah, the oldest mining area in the United States, and mines in the Arizona Strip district in Arizona must ship their ore to White Mesa for processing. Development of RAR of more than 14,000 tU on the Colorado Plateau and of RAR of 2,500 tU in the Arizona Strip is strongly influenced by whether this ore will have access to economic milling.

Exploitation of the largest uranium deposit in the world in Australia, BHP Billiton's Olympic Dam in Australia , depends on the development of extensive infrastructure, including building a desalination plant at the Indian Ocean and piping water approximately 300 kilometers (km) (186 miles (mi)) to the mine site in order to provide adequate water to expand the mine. The 2012 decision to postpone development of this resource (ABC News, 2012) illustrates how strongly resource development depends on adequate infrastructure.

Largest Corporate Entities and Holdings

Approximately 25 companies are producing uranium worldwide. Of these companies, the top fourteen provided 91 percent of mined uranium in 2010. Major world uranium producers are Areva NC (Areva), the Cameco Corporation (Cameco), KazAtomProm , Rio Tinto, Atomredmetzoloty OJSC (ARMZ), Navoi Mining and Metallurgy Combinat (Navoi), BHP Billiton, Paladin Energy Ltd. (Paladin), and Uranium One Inc. (Uranium One) (Ux Consulting LLC, 2010; World Nuclear Association, 2011b). Distribution of uranium production among the top companies in 2010 is shown in figure 16.

Ux Consulting LLC predicts that it is likely that Areva, ARMZ (now ARMZ/Uranium One), BHP Billiton, Cameco, KazAtomProm and Rio Tinto will maintain a large market share of production to 2020. Projections show that Areva's production will remain relatively constant at about 20 percent of world supply. Cameco's share will fall slowly through 2020 to about 13 percent. KazAtomProm is expected to remain relatively flat, falling slightly to 12 percent by 2020, and ARMZ is expected to fall significantly to 5 percent by 2020, although their 2010 acquisition of Uranium One (which is projected to have 6 percent of the world market by 2020) will bolster production. Rio Tinto is expected to fall in share to less than 10 percent, and BHP Billiton is expected to fall to 5 percent.

Areva has the largest geologically defined resource base. Next in that category are BHP Billiton, KazAtomProm and Cameco (fig. 17). Rio Tinto is mining resources at

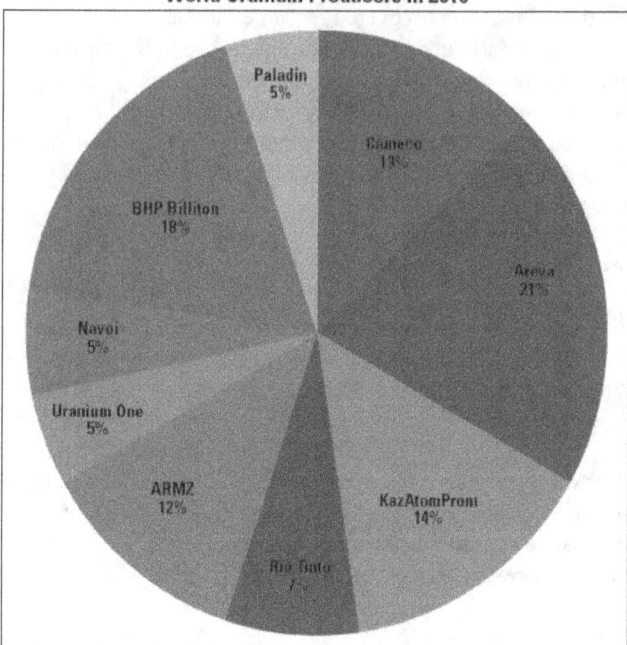

Distribution of Uranium Resources Among the Top Nine World Uranium Producers in 2010

Figure 17.   Distribution of resources among the top 9 world uranium producers in 2010. Data from Ux Consulting Company LLC (2010).

the most rapid pace, followed by Cameco, Uranium One, KazAtomProm, Navoi, Areva, Paladin, and ARMZ. The rapid rate at which Rio Tinto is depleting its resource is related to the ages of its two producers, Ranger and Rossing, both of which are reaching the end of their effective mine lives. The two large ISL producers in the world market, KazAtomProm and Uranium One, are depleting RAR at a relatively rapid pace due to the nature of ISL mining, which can be brought online quickly and operates for a relatively short period of time. Cameco's production relies heavily on their properties in the Athabasca Basin of Canada, where uranium can be mined rapidly due to the high-grade nature of the ore. BHP Billiton's Olympic Dam mine is being depleted at a low rate because it is a large, low-grade deposit.

Long-term Demand and Supply Projections

Projected Production of Supply, by Country

Uranium production from Kazakhstan is expected to increase in response to the aggressive marketing and development of resources by KazAtomProm, the short lead-time for development, the low cost of infrastructure required for ISL mines, and this country's large economic RAR (fig. 18, 19).

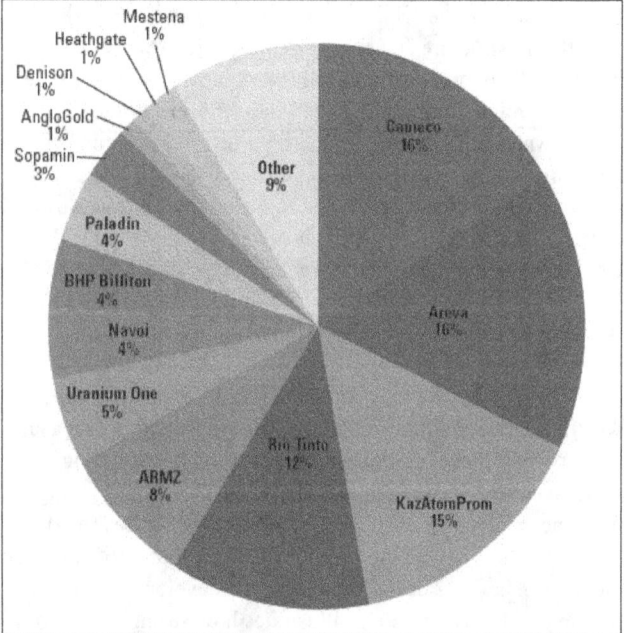

2010 World Production by Company

Figure 16.   Uranium production by company in 2010. Data from the World Nuclear Association, (2011b).

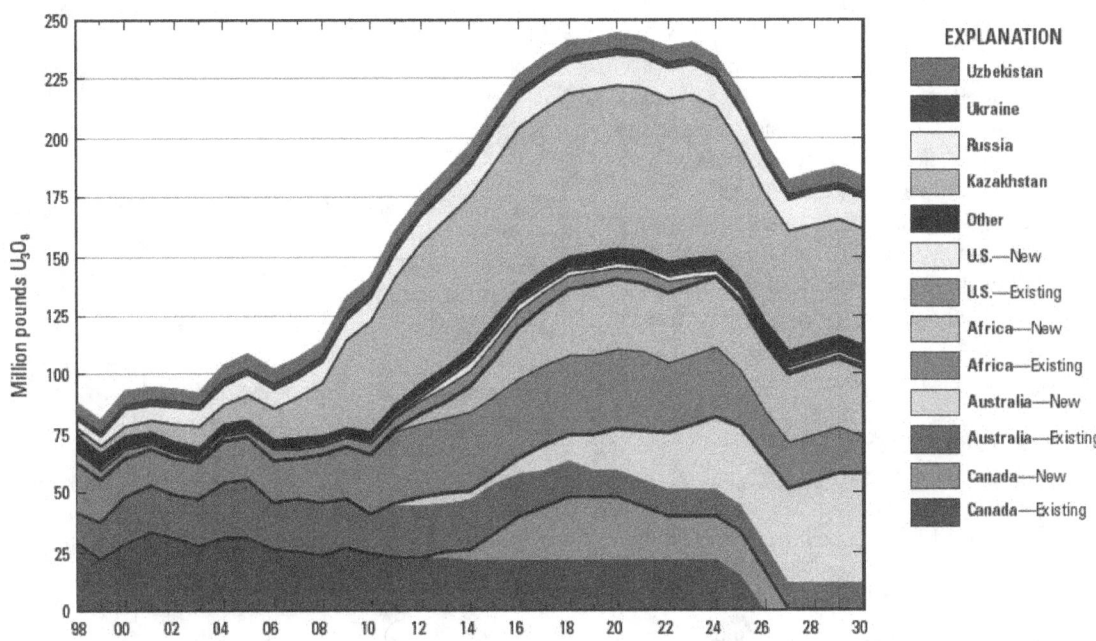

Figure 18. Existing uranium production and projected trends in production worldwide, 1998–2030. From Ux Consulting Company LLC (2010), reproduced with permission.

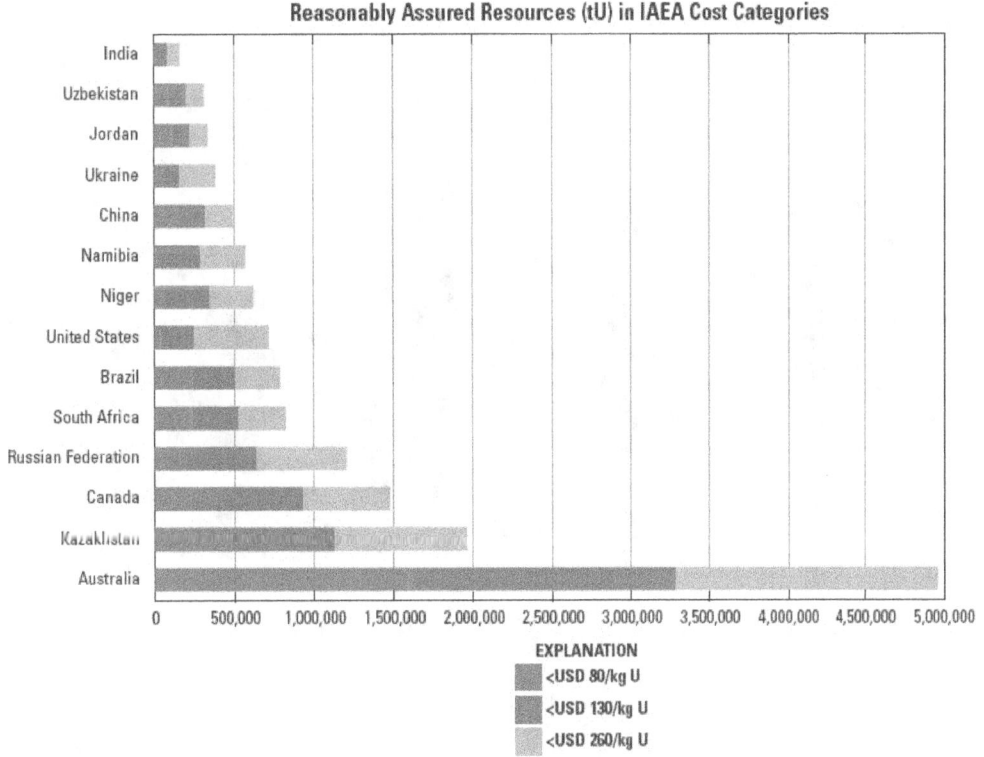

Figure 19. Distribution of reasonably assured resources (RAR) in Nuclear Energy Agency (NEA) cost categories for the most resource-rich countries. Data from NEA–IAEA (2010).

Fourteen operating mines should continue to produce, joined by two new mines. The rate at which production increases in Kazakhstan depends, in part, on the rate at which the regional infrastructure develops and recent rapid increases may not be sustainable in the future. In Australia, existing deposits, such as Ranger, should produce significantly less uranium, but large new properties, such as Yeelerie, Olympic Dam, Ranger Deeps, and the smaller ISL properties–Four Mile and Honeymoon–may more than make up for mines that are closing, or for those whose production is expected to decrease. New deposits from Africa, in particular from Niger (Imouraren) and from Namibia (Trekkopje), could bring a great deal of production online. Namibia's production from existing mines (notably, in Langer Heinrich and in Rossing) continues. In Canada, some of the largest producing mines may be depleted (McArthur River, Rabbit Lake), and, although new planned mines (Midwest, Cigar lake) should make up for this production, their contribution to world supply is predicted to be relatively short-lived. Production from other countries, including Uzbekistan, Russia, Ukraine, and others, is expected to continue steadily into the future. Uranium production in the United States is minor, providing less than 10 percent of domestic demand, and properties now being developed are relatively small ISL mines that are fairly short lived (Ux Consulting Company LLC, 2011).

Continued growth in uranium production worldwide depends in part on sustained prices that support extraction of resources in the higher cost categories (fig. 19).

Projections to 2035

On the basis of data supplied from member countries, the NEA reports that growth in global installed nuclear capacity is projected to grow from 372 gigawatts of electricity (GWe) net in early 2009 to between 511 GWe net (low case) and 782 GWe net (high case) by the end of 2035 (NEA–IAEA, 2010). Projected uranium requirements vary considerably by region (fig. 20). Most of the projected growth is in East Asia, where increases in annual uranium requirements are forecast to range from 120 percent to 180 percent more than 2009 requirements. By contrast, annual requirements in North America are expected to range from a 25-percent decrease in the low case to a 55-percent increase in the high case. The European Union's requirements are projected to range from a 15-percent decline to an increase of more than 25 percent by 2035.

Projected nuclear capacity translates to increased global reactor-related uranium requirements that range from 227 million pounds U_3O_8 per year (87,370 tU/yr) to 359 million pounds U_3O_8 per year (138,165 tU/yr), assuming a tails assay

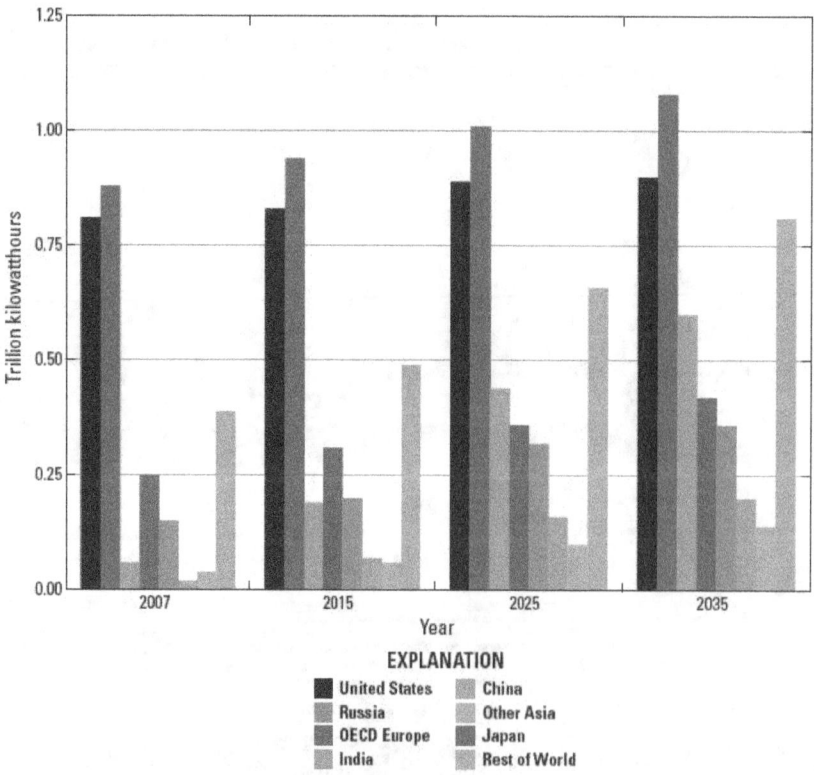

Figure 20.   World net electricity generation from nuclear power, by region, 2007–2030 (Energy Information Administration, 2010b).

of 0.30 percent, by the end of 2035 (fig. 21). This represents an increased uranium requirement of 40 percent in the low case, and 120 percent in the high case (NEA–IAEA, 2010).

Primary production of uranium, that is, actual mining of uranium, accounted for 50 to 60 percent of world requirements during 2006–2011, with secondary sources providing the remainder. Secondary sources, especially the source that will diminish after 2013 when the United States/Russian program, Megatons to Megawatts, expires, will require an increase in uranium from other sources in the near term to meet future demand (NEA–IAEA, 2010).

Sixty percent of the 3.5 million tU mineable for less than USD 130/kgU identified worldwide is either in production, or scheduled to come online by 2015. If uranium prices trend higher, an additional 479,600 tU identified in the higher cost categories worldwide should become economic (fig. 13). Mining of inferred resources reported by NEA—those classified as less well constrained geologically—could add another 2.3 million tU of economic resources to world supply (NEA–IAEA, 2010). With no growth in demand, these resources would fulfill current requirements: inferred resources for an additional 27 years, and currently uneconomic resources for an additional 6 years. Coupled with demonstrated resources, NEA's reporting predicts that RAR in the identified and in the inferred

categories would satisfy current demand for 70 years. However, the inferred categories are poorly understood and could be substantially different from current estimates.

NEA estimates that, as of 2009, existing and committed uranium mine production covers global demand through 2021 in the high-case, and until 2024 in the low-case growth scenarios. Assuming that plans are successful for significant expansion of existing mines and for new production centers, high-case demand requirements will be met until 2029, and low-case requirements until after 2035 (NEA–IAEA, 2010).

The Ux Consulting Company LLC publishes projections of uranium supply determined by detailed examination of development timelines for existing mines and for processing facilities, for planned and for potential mines, and for secondary supplies. Their projections indicate that supply will increase 13 percent to 159 million pounds U_3O_8 (61,158 tU) by 2011, and that the increase will range from 210 to 224 million pounds U_3O_8 (from 80,775 to 86,160 tU) by 2020. In any scenario for nuclear growth, future projections based on high- and on low-growth scenarios show that secondary uranium supplies are required to fuel the world nuclear power reactor fleet (fig. 22). Projections by both NEA and the Ux Consulting Company indicate that mine development is proceeding too slowly to fully meet requirements for an expanded nuclear power reactor fleet, and imbalances in supply and demand may occur.

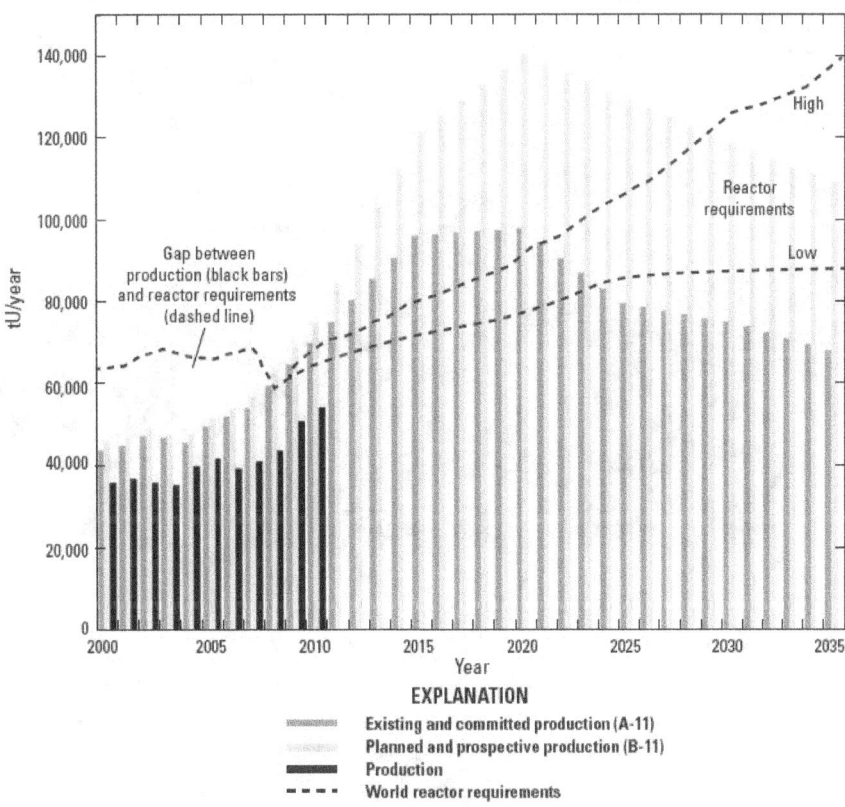

Figure 21.   Scenarios for production of and requirements for uranium, 2011–2035. From NEA–IAEA (2010), reproduced with permisison.

Projections beyond 2035

The World Nuclear Association (WNA), an international private-sector organization, is one of the few organizations that projects global demand for nuclear power to 2100. Figure 23 illustrates the WNA upper and lower trajectories for growth of nuclear power. The WNA provides an analysis of upper and lower trajectories for each country that has a nuclear program. The lower trajectory reflects the minimum anticipated nuclear capacity of ~2,050 GW, and the upper trajectory reflects a full policy commitment to nuclear power that is as much as 11,000 GW of capacity. The zone between the upper and lower trajectories is considered the likely domain in which growth will occur, and the trajectories are considered boundaries rather than low or high cases. The WNA takes the optimistic view that uranium supply will not be an obstacle to future growth, and that the combined factors of successful exploration, new mining technologies, reprocessing of fuel, and use of Generation IV reactors, will ensure an ample supply of nuclear fuel well into the 22nd century. WNA does concede that delays in bringing RAR into production are increasingly challenging (World Nuclear Association, 2011a).

The NEA takes a more cautious view of the timely development of nuclear fuel supplies. In the 2009 "Red Book", the NEA forecasts adequate Identified Resources (Reasonably Assured Resources plus Inferred Resources) to supply reactors for the next 100 years, if 2008 consumption rates (154 million pounds U_3O_8 per year (59,065 tU/yr)) are projected into the future. This forecast does not take into account projected growth in capacity (NEA–IAEA, 2010). If all conventional resources are included (Identified Resources plus Speculative and Prognosticated Resources), then the supply would last 300 years, through 2410, using the 2008 consumption rate. In the 2009 "Red Book" and in related publications, the NEA emphasizes that potential problems in the supply chain do not include the resource base, but instead lie with factors that affect timely development of RAR. For example, expanding the resource through exploration and through development depends on a robust market that also provides the required capital and financing to bring mines online. Other factors that influence uranium resource development include the region's regulatory climate for mining, its established regulations and safeguards for safe mining development and operations, an adequate infrastructure, a skilled workforce, and the region's ability to retain

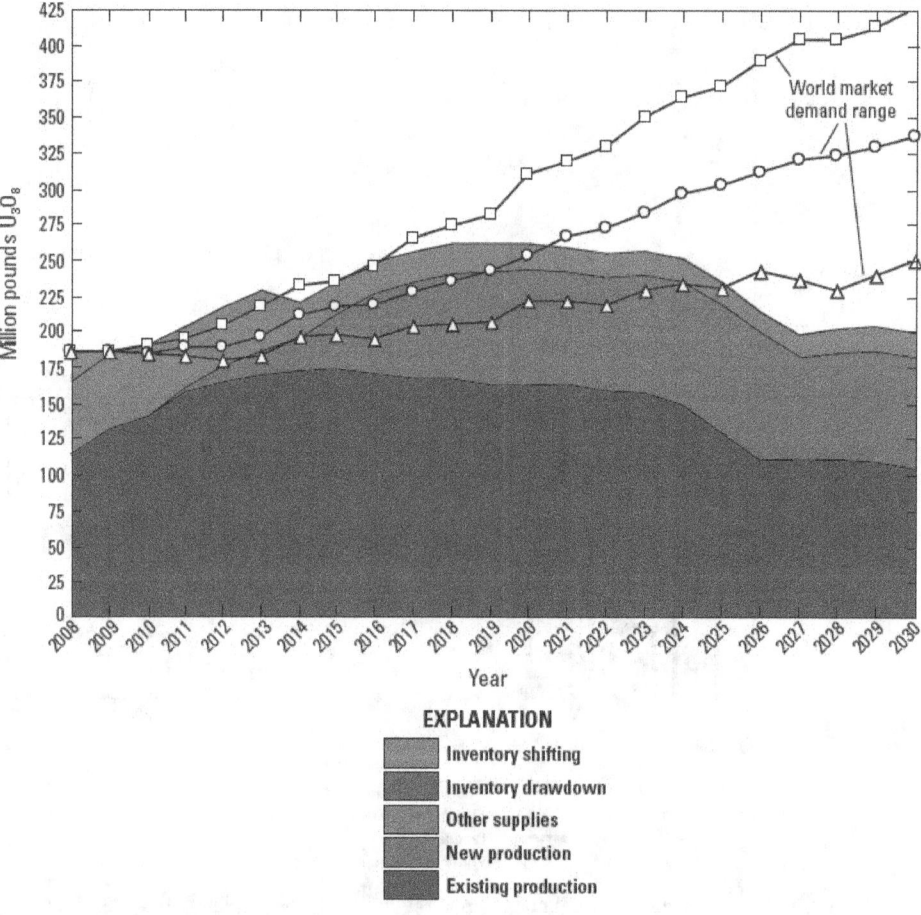

Figure 22. Actual and projected world trends in growth of market demand and increase of supply, 2008–2030. From Ux Consulting Company LLC, (2010), reproduced with permission.

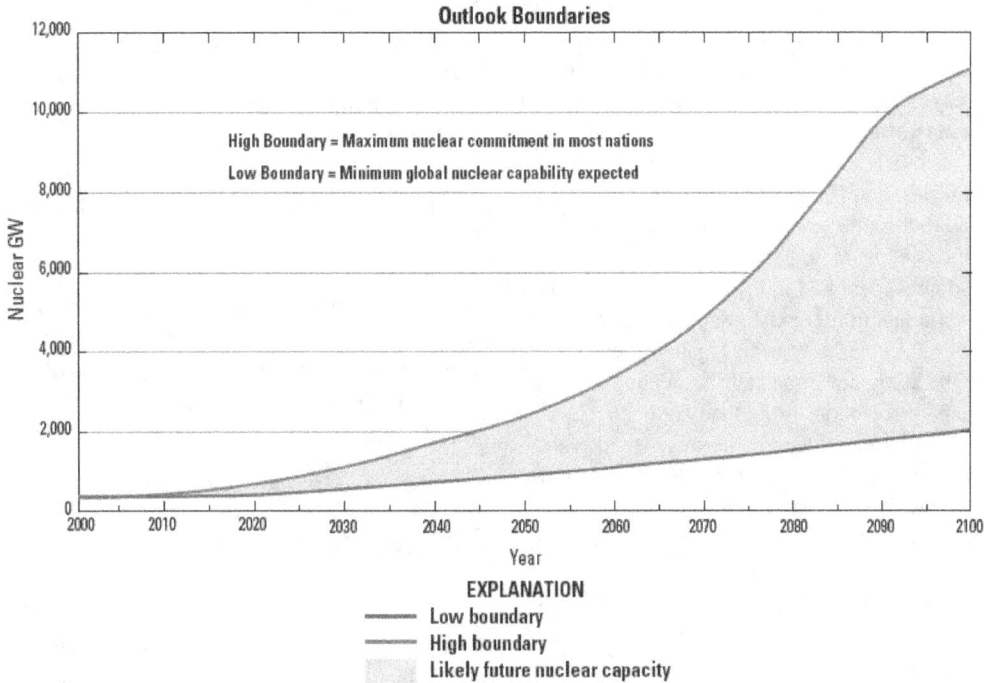

Figure 23. World Nuclear Association projections for future nuclear capacity worldwide, to 2100. From World Nuclear Association (2011b), reproduced with permission.

financiers willing to wait many years for a return on their investment. The single most important factor identified for ensuring that mines are developed in a timely fashion is the strength of the market. In uncertain financial times, a sustained and strong uranium market is far from guaranteed.

Most projections of uranium supply and demand focus on primary production from mines that extract uranium as a primary product or byproduct, and on secondary sources to fulfill demand. Although higher uranium costs may stimulate exploration and new discoveries, projections to 2100 and beyond would not be complete without evaluating the potential uranium contribution from unconventional resources such as uraniferous phosphates, black shales, and seawater. Uranium is elevated significantly above crustal abundance in both black shale and in phosphate deposits, and the economic extraction of uranium from these sources is being actively investigated. The extraction of uranium from phosphates is receiving the most attention, because it can potentially tap into a vast resource that may exceed 1.9 billion tU (IAEA, 2010). The environmental benefit of removing uranium from phosphates that are primarily mined for fertilizer makes this potential uranium source increasingly attractive. Uranium in black shales is also a large resource capable of yielding an estimated 1 billion tU. The benefit of developing techniques for removing uranium from existing polymetallic mines has become the focus of some large uranium producers. Uranium extraction from seawater is still in the research stage, and is currently not even close to being economic, with reported production costs reaching USD 700/kgU (NEA–IAEA, 2010).

Summary

This report addresses the question: Is there enough uranium in the worldwide resource base to meet the cumulative worldwide requirement for the decades through 2035? Much of the emphasis of the report rests therefore on the physical quantities of uranium resources, including a fairly detailed comparison of what is known about supply and what is known about likely demand. The report concludes that identified uranium resources appear to be adequate to serve needs into the 25-year future, but that delays in developing these resources may create periodic imbalances.

Mine production may not keep pace with demand because (1) the time between the delineation of a deposit and the time when it is first mined can lag by as much as 15 years ; (2) exploration in some regions is insufficient to keep production growing at reasonable rates; (3) infrastructure inadequate to support the economic milling of the ore may limit extraction; and (or) (4) future exploratory drilling may reveal less resource than is currently estimated, especially in the categories of mines that are less geologically certain.

It is also possible, however, that the report has erred in the direction of being too conservative. Only a few of the many technological changes and substitutions that are likely to occur on the supply and the demand sides of the market have been incorporated into mining procedures. In many cases, additional resources are discovered as mining proceeds. Although this report describes a few cases where uranium mining uncovered much larger resources nearby, it would be too speculative to include this kind of growth in estimates of reserve.

In the short term, an analysis of production centers shows that, since 1990, uranium has been under produced when compared to demand. The shortfall has been made up from stockpiles, from recycling, and from the United States/Russian HEU–LEU program (to end in 2013). Although worldwide reporting restrictions make it difficult to determine remaining uranium inventories, it appears that these stockpiles are being depleted, and that primary supply is coming online more slowly than demand is expanding.

Although resources identified in the existing and developing supply are sufficient to fuel existing reactors at current rates of consumption for at least 30 years (to 2040), capacity does not appear to be being developed at a fast enough rate to keep up with demand. Projections more than about 20 years into the future (to 2030) are problematic because whether uranium resources will be identified depends on the price of uranium, which in turn drives both exploration and development. If, however, we assume that NEA's estimates of world supply are accurate, so that higher-cost uranium deposits and those that are not geologically well constrained (inferred resources) are included in estimates of supply, then, at current rates of consumption, supply can keep pace with demand for 70 years (to 2080). This evidence-based conclusion does not take into account the future growth worldwide of nuclear power.

Predictions that show expansion of nuclear capacity as ranging from 40 to 120 percent of current capacity by 2035 will strain capacity for developing supply. NEA estimates that operating and developing mines will be capable of satisfying only 80 percent of demand for the low-growth scenarios, and 50 percent of demand for the high-growth scenarios. Recognizing this potential shortfall, some utilities are buying into primary supply at the mine site rather than relying on uranium suppliers or brokers.

Additional concerns include the possibility that the stability of future primary uranium supplies will decrease. More primary uranium will be supplied from Kazakhstan, Africa (Namibia, Niger), Australia, and Canada, with production from other countries remaining flat. Production in Australia is tied to the large Olympic Dam deposit, and Canada largely depends on the development of the Cigar Lake and the Midwest mines. The dependence of uranium supply on large individual uranium properties and countries adds uncertainty to estimates of future supply. Major producers Cameco, Areva, KazAtomProm, Rio Tinto, ARMZ/Uranium One, and BHP Billiton are expected to continue to maintain their large market share into the future.

Growth in nuclear capacity worldwide will put pressure on existing and identified supplies. Unless new large-capacity mines come online in the near future, prices are expected to rise, and this increase should at the same time stimulate additional exploration and make some unconventional resources more attractive. Long-term projections of uranium supply depend on the rate at which nuclear capacity expands.

As supplies tighten, utilization of secondary sources and of unconventional resources is likely to expand. Industry is beginning to develop some large mostly unexploited resources, including uranium in phosphates and in black shale deposits. In addition, secondary sources including uranium in enrichment tails are increasingly attractive. Exploration is expected to continue if uranium prices remain high, moving the reported 2.81 million tU of prognosticated, and the 7.5 million tU of speculative undiscovered resources into less hypothetical RAR categories. Although data are currently insufficient to allow accurate projections of the extent to which unconventional resources will contribute to the expansion of uranium supply into the future, careful monitoring of ongoing pilot projects and of expansion of capacity is warranted.

Acknowledgments

This paper benefited from reviews by and discussion with Robert Vance of the OECD Nuclear Energy Agency, and William Watson and Scott Sitzer (retired) of the DOE Energy Information Administration. The authors would also like to acknowledge Mark Hannon, who drafted the maps in appendix 1.

References Cited

ABC News, 2012, BHP shelves Olympic Dam as profit falls a third, August 22, accessed September 24, 2012 at: *http://www.abc.net.au/news/2012-08-22/bhp-billiton-profit-falls-a-third/4215638*.

Areva NC, 2011, [Home page]: Areva Web site, accessed January 2011, at *http://www.areva.com*.

Bakarzhiyev, Y.A., 2011, Recent developments in the field of uranium—Ukraine: Joint Nuclear Energy Agency/ International Atomic Energy Agency Group on Uranium, meeting, 47, Paris, France, November 2–4, 2011, presentation, 14 p.

Berkeley Resources Ltd., 2011, Home: Berkeley Resources Ltd. Web site, accessed January 2011, at *http://www.berkeleyresources.com.au/*.

BHP Billiton, 2011, Uranium: BHP Billiton Web site, accessed January, 2011, at *http://www.bhpbilliton.com/*.

Bianchi, R.E., 2010, Update of raw materials activities in Argentina: Nuclear Energy Agency-International Atomic Energy Agency Group on Uranium, meeting, 45, Saskatoon, Canada, August 19–20, 2010, presentation, 29 p.

Boytsov, A.V., 2010, Sustainable development of U production—Time challenge: Joint Nuclear Energy Agency/ International Atomic Energy Agency Group on Uranium, technical meeting, Saskatoon, Canada, 15 p.

Calvert, Tom, 2010, Recent developments—Canada: Joint Nuclear Energy Agency/ International Atomic Energy Agency Group on Uranium, meeting, 45, Saskatoon, Canada, August 19–20, 2010, presentation, Organisation for Economic Co-operation and Development Nuclear Energy Agency and International Atomic Energy Agency, 15 p.

Cameco Corporation, 2011, On the double—Keeping pace with uranium demand: Cameco Corporation 2010 annual report, 144 p. (Also available at *http://www.cameco.com/ investors/financial_reporting/annual_reports/2010/*.)

Chaki, Anjan, 2010, India country report: Joint Nuclear Energy Agency/ International Atomic Energy Agency Group on Uranium, meeting, 45, Saskatoon, Canada, August 19–20, 2010, presentation, 16 p.

Commonwealth of Australia, 2010, Australia's in situ recovery uranium mining best practice guide—Groundwaters, residues and radiation protection: [Australia] Department of Resources, Energy and Tourism, 23 p. (Also available at *http://www.ret.gov.au/resources/Documents/ Mining/10-4607_in%20situ%20uranium_31May.pdf*).

Dahlkamp, F.J., 2010, Uranium deposits of the world: Berlin, Springer, v. 4, 2,400 p.

daSilva, L.F., 2010, Brazil, recent activities: Joint Nuclear Energy Agency/ International Atomic Energy Agency Group on Uranium, meeting, 45, Saskatoon, Canada, August 19–20, 2010, presentation Organisation for Economic Co-operation and Development–Nuclear Energy Agency & International Atomic Energy Agency, 2 p.

Energy Fuels Inc., 2012, Home page: Energy Fuels Inc. Web site, accessed October, 2012 at *http://www.energyfuels.com/*.

Energy Information Administration, 2009, Domestic uranium production report—Annual [2009]: U.S. Energy Information Administration, accessed January 2011 at *http://www.eia.gov/cneaf/nuclear/dupr/dupr.html*.

Energy Information Administration, 2010a, Domestic uranium production report—Quarterly, 4th quarter 2010: U.S. Energy Information Administration, accessed March 2011, at *http://www.eia.gov/cneaf/nuclear/dupr/qupd.html*.

Energy Information Administration, 2010b, International energy outlook 2010 with projections to 2035: U.S. Energy Information Administration Report no. EIA–0554, 195 p.

Energy Information Administration, 2010c, International energy statistics: U.S. Energy Information Administration database, accessed March 2011 at *http://205.254.135.7/ cfapps/ipdbproject/iedindex3.cfm*.

Energy Information Administration, 2010d, Uranium marketing annual report [2009]: U.S. Energy Information Administration, accessed April, 2010 at *http://www.eia.gov/cneaf/ nuclear/umar/umar.html*.

Energy Information Administration, 2011, Domestic uranium production report 2010: U.S. Energy Information Administration, 14 p., accessed May 2011 at *http://205.254.135.7/ uranium/production/annual/*.

Greenland Minerals and Energy Ltd., 2011, [Home page]: Greenland Minerals and Energy Ltd. Web site, accessed January 2011, at *http://www.ggg.gl/*.

Hisatani, Koichi, 2010, Current status of nuclear industry and technology in Japan: Joint Nuclear Energy Agency/ International Atomic Energy Agency Group on Uranium, meeting, 45, Saskatoon, Canada, August 19–20, 2010, presentation, 7 p.

IAEA (International Atomic Energy Agency), 2009, World distribution of uranium deposits (UDEPO) with uranium deposit classification: Vienna, Austria, International Atomic Energy Agency IAEA–TECDOC–1629, 117 p.

IAEA (International Atomic Energy Agency), 2010, World distribution of uranium deposits (UDEPO): International Atomic Energy Agency database, accessed January–April, 2011, at *http://infcis.iaea.org/*.

Itamba, Helena, 2011, Developments in the field of uranium Namibia: Joint Nuclear Energy Agency/International Atomic Energy Agency Group on Uranium, meeting, 45, Saskatoon, Canada, August 19–20, 2010, presentation, 14 p.

Jones, Bryn, and Davidson, James, 2009, PhosEnergy—New age extraction of uranium from phosphoric acid, *in* International Atomic Energy Agency Uranium from unconventional resources technical meeting, August 2009, Vienna, Austria, International Atomic Energy Agency, 33 p.

Khan Resources Inc., 2011, Latest news: Khan Resources Inc. Web site, accessed April 2011, at *http://www.khanresources.com/*.

Marenica Energy Limited, 2011, Marenica Energy Limited is an Australian Stock Exchange listed company with its major focus being the discovery and development of uranium deposits: Marencia Energy Limited, accessed April 2011, at *http://www.marenicaenergy.com.au/*.

Mawson Resources Limited, 2008, Tåsjö uranium project: Mawson Resources Limited, accessed February 2011, at *http://www.mawsonresources.com/s/Tasjo.asp*

McKay, Aden, and Carson, Leesa, 2010, Recent developments in Australia's uranium mining industry: Joint Nuclear Energy Agency/International Atomic Energy Agency Group on Uranium, meeting, 45, Saskatoon, Canada, August 19–20, 2010, presentation, 28 p.

Mega Uranium Mining Company, 2011, Uranium resources: Mega Uranium Mining Company, accessed February 2011, at *http://www.megauranium.com/uranium_resources/*.

MIT Energy Initiative, 2011, The future of the nuclear fuel cycle: Massachusetts Institute of Technology, 237 p. (Also available at *http://web.mit.edu/mitei/research/studies/nuclear-fuel-cycle.shtml*.)

NEA–IAEA (Organization for Economic Co-operation and Development Nuclear Energy Agency and International Atomic Energy Agency), 2000, Uranium 1999—Resources, production and demand: Organization for Economic Co-operation and Development Publishing, 340 p.

NEA–IAEA (Organization for Economic Co-operation and Development Nuclear Energy Agency and International Atomic Energy Agency), 2006, Uranium 2005—Resources, production and demand: Organization for Economic Co-operation and Development Publishing, 388 p.

NEA–IAEA (Organization for Economic Co-operation and Development Nuclear Energy Agency and International Atomic Energy Agency), 2008, Uranium 2007—Resources, production and demand: Organization for Economic Co-operation and Development Publishing, 420 p.

NEA–IAEA (Organization for Economic Co-operation and Development Nuclear Energy Agency and International Atomic Energy Agency), 2010, Uranium 2009—Resources, production and demand, Organization for Economic Co-operation and Development Publishing, 456 p.

NECSA (South African Nuclear Energy Corporation), 2010, Uranium in South Africa—An industry update: Joint Nuclear Energy Agency/ International Atomic Energy Agency Uranium Group, meeting, 45, Saskatoon, Canada, August 19–20, 2010, presentation, 12 p.

Ngoupana, Paul-Marin, Felix, Bate, and Barker, Anthony, 2011, Areva suspends CAR uranium mine project: Thompson Reuters, November 3, accessed June 22, 2012, at *http://af.reuters.com/article/commoditiesNews/idAFL5E7M34T920111103*.

Nuclear Energy Agency (NEA), 2006, Forty years of uranium resources, production and demand in perspective: Organization for Economic Co-operation and Development Publishing, 276 p.

Nuclear Energy Agency (NEA), 2008, Nuclear energy outlook 2008: Organization for Economic Cooperation and Development Publishing, 460 p.

Paladin Energy, Ltd., 2011, Paladin Energy Ltd.—The new energy on the market: Paladin Energy, Ltd. Web site, accessed June 22, 2012, at *http://www.paladinenergy.com.au/*.

Pool, T.C., and Ross, D.A., 2007, Technical report on the Arizona strip uranium project, Arizona, U.S.A.: Scott Wilson Roscoe Postle Associates Inc. technical report NI 43–101, prepared for Denison Mines Corporation, 116 p., accessed June 22, 2012, at *http://www.denisonmines.com/content/pdf/43-101_arizona_strip.pdf*.

Puritch, Eugene, Hayden, Alfred, Partsch, Alexander, Harron, G.A., and Brown, F.H., 2010, Preliminary economic assessment on the Viken MMS project, Sweden: P & E Mining Consultants Inc., EHA Engineering Ltd., and G.A. Harron & Associates Inc. technical report NI 43–101 and 43–101F, prepared for Continental Precious Minerals Inc., 122 p., accessed June 22, 2012, at *http://www.czqminerals.com/assets/pdf/viken_pea_report_oct_2010.pdf*.

Rigby, Neal, Muller, S.C., Hollenbeck, Patrick, Stryhas, Bart, Daviess, Frank, and Kurrus, Andy, 2010, Technical report on resources—Uranium Energy Corp. Palangana ISR uranium project, deposits PA–1, PA–2 and adjacent exploration areas, Duval County, Texas: SRK Consulting Engineers and Scientists technical report NI 43–101, prepared for Uranium Energy Corp., 106 p., accessed June 22, 2012, at *http://www.uraniumenergy.com/_resources/reports/Palangana_NI_43-101_Technical_Report_199600_010_KG_012-opt.pdf*.

Spiering, E.D., Hillard, P.D., and Inman, J.R., 2009, Exploration and discovery of blind breccia pipes—The potential significance to the uranium endowment of the Arizona strip district, northern Arizona: U2009 Global Uranium Symposium, Keystone, Colorado, May 9–13, 2009, abstracts, p. 17.

Talvivaara Mining Company Plc., 2011, Mineral resources: Talvivarra Mining Company Plc. Web site, accessed June 22, 2012, at *http://www.talvivaara.com/operations/Mineral_resources*.

Titley, Malcolm, 2009, The Mutanga project—Located in Southern Province, Republic of Zambia: GSA Global (UK) Ltd. technical report NI 43–101, prepared for Denison Mines Corp., 225 p., accessed June 12, 2012, at *http://www.denisonmines.com/content/pdf/ni43_101_mutanga_uranium_19mar09.pdf*.

Toro Energy Ltd., 2011, Projects: Toro Energy Limited Web site, accessed June 12, 2012, at *http://www.toroenergy.com.au/projects.html*.

TradeTech, 2011, TradeTech—Specialists in nuclear fuel markets—Consulting and supply/demand analysis: Tradetech Web site, accessed [date], at *http://www.uranium.info/*.

U.S. Enrichment Corporation Inc., 2011, Megatons to Megawatts: USEC Inc. Web site, accessed February 2011, at *http://www.usec.com/megatonstomegawatts.htm*.

Ux Consulting Company LLC, The, 2010, Uranium suppliers annual: The Ux Consulting Company, LLC special report, 500 p.

Ux Consulting Company LLC, The, 2011, The industry's leading source of consulting, data services & publications on the global nuclear fuel cycle markets: The Ux Consulting Company, LLC Web site, accessed January, 2011, at *http://www.uxc.com/index.aspx*.

Vance, Robert, 2005, What can forty years of red books tell us?: Organization for Economic Co-operation and Development Publishing, 19 p.

Vettraino, Fortunato, 2010, Brief update on the nuclear energy programme in Italy and related uranium supply features: Joint Nuclear Energy Agency/International Atomic Energy Agency Group on Uranium, meeting, 45, Saskatoon, Canada, August 19–20, 2010, presentation, Organisation for Economic Co-operation and Development Nuclear Energy Agency and International Atomic Energy Agency 6 p.

Vettraino, Fortunato, 2011, Updates to the nuclear energy programme in Italy and related uranium outlook: Joint Nuclear Energy Agency/International Atomic Energy Agency Group on Uranium, meeting, 47, Issy Les Moulineaux-Paris, France, November 2–4, 2011, presentation, 7 p.

World Nuclear Association, 2009, Projected uranium requirements to 2030: World Nuclear Association Web site, accessed January 2011, at *http://www.world-nuclear.org/*.

World Nuclear Association, 2011a, The global nuclear fuel market; Supply and demand, 2011 to 2030: World Nuclear Association, 197 p.

World Nuclear Association, 2011b, [Home page]: World Nuclear Association Web site, accessed January 2011, at *http://www.world-nuclear.org/*.

World Nuclear News, 2007, New study on uranium from phosphates: World Nuclear News, July 31, accessed March, 2011, at *http://www.world-nuclear-news.org/newsarticle.aspx?id=13796&LangType=2057*.

Zhang, Decun, 2010, Uranium resources, production and requirements in China: Joint Nuclear Energy Agency/International Atomic Energy Agency Group on Uranium, meeting, 45, Saskatoon, Canada, August 19–20, 2010, presentation, 8 p.

Appendix 1.   Analysis of Operating Mines and Advanced-Stage Uranium Properties, by Country

This appendix describes uranium RAR and production in the most important countries that contain uranium resources. In addition to uranium that is actively being produced or is in potential near-future production, the country narratives discuss projects that are likely to take longer to develop than those studied for this report. Possible impediments to or potential increases in uranium production are discussed for each major uranium-producing country worldwide.

Argentina

Argentina reports 10,400 tU mineable in the top two cost categories (<USD 130/kgU and <USD 260/kgU), and it reports 7,000 tU mineable at less than USD 80/kgU. Argentina mined uranium for use in its domestic reactors until the late 1990s when it suspended domestic mining because less expensive uranium available on the world market became more attractive. In 2006, national policy tasked the Comision Nacional de Energia Atomica (CNEA) to restart local production, and government exploration expenditures have increased dramatically (Bianchi, 2010; NEA–IAEA, 2010). Two mines—the Sierra Pintada deposit (2,620 tU), part of the San Rafael Mining and Milling complex, and Cerro Solo (3,900 tU)—are on standby while CNEA evaluates the feasibility of restarting the mining of these deposits (IAEA, 2009; Bianchi, 2010; Ux Consulting Company LLC, 2010) (fig. 1–1). Argentina has also begun a program to assess the country's potential undiscovered resource.

Australia

The three operating uranium mines in Australia include two of the most prolific world producers during 2009. The large underground Olympic Dam (Roxby Downs) mine, operated by BHP Billiton, produced 2,981 tU, and the open-pit Ranger mine operated by Rio Tinto produced 4,423 tU. Heathgate Resources Pty. Ltd (Heathgate) owns and operates the Beverly ISL mine (South Australia), which produced 559 tU (McKay and Carson, 2010) (fig. 1–2).

Olympic Dam is a polymetallic Iron Oxide Copper Gold deposit currently mined primarily for gold and copper, with uranium production (capacity 3,820 tU/yr) incidental to these commodities. The total RAR (295,000 tU) ranks Olympic Dam as the world's largest uranium deposit, holding 34 percent of the world's total identified uranium resource (Mckay and Carson, 2010). Although uranium's economic contribution to the mine's economic viability is significant, its low grade

Figure 1–1.   Location of uranium mines in Argentina, 2010. From Bianchi (2010).

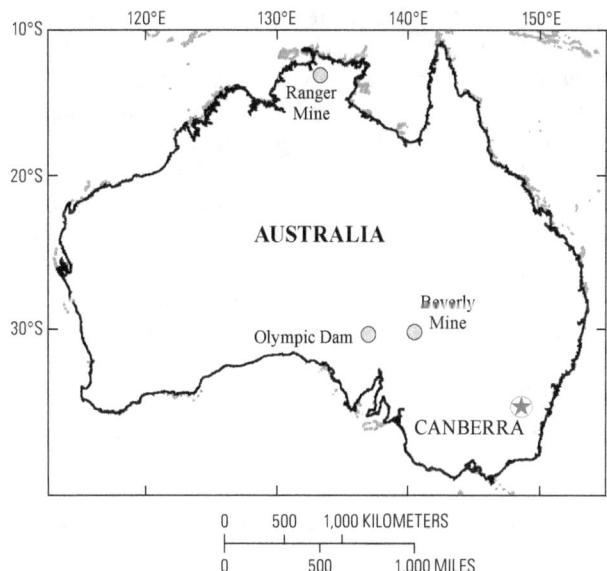

Figure 1–2.   Location of operating mines producing uranium in Australia. From Mckay and Carson (2010).

(0.023 percent U) would, at current prices, probably preclude production of uranium alone from this mine. Olympic Dam is, however, the world's fourth largest remaining copper deposit, and the fifth largest gold deposit. The future price of copper and gold will therefore more probably influence uranium production from Olympic Dam. Technical issues have interrupted production of uranium at Olympic Dam in the past. Most recently, production was cut 75 percent during October 2009 through March 2010 because a hoist that transported ore to the surface failed.

A proposed expansion project that will increase the uranium production capacity at Olympic Dam to 16,100 tU is in the permitting process (IAEA, 2009). A feasibility study is complete, and a draft Environmental Impact Statement has received public comment and is in review by the government. This expansion includes the construction of a large openpit mine estimated to eventually be 6 km by 5 km and more than 1,200 meters deep, southeast of the current underground mine. The cost of this expansion will range from 30 to 50 billion U.S. dollars. One of the greatest technical challenges for the project is securing adequate water to run the proposed operation. BHP Billiton's preferred solution is to desalinate ocean water at Point Lowly in the Upper Spencer Gulf and pipe the water 320 km (200 mi) to the mine. While this paper was in review, BHP Billiton announced it would suspend expansion of this pit, pushing production of this uranium to an undetermined time in the future (ABC News, 2012). This supply was projected to have become available in about 2020, and provided significant primary supply to the world market. The impact of the loss or delay of this source on forward supply projections for uranium has not been fully evaluated.

The Ranger openpit mine (28,832 tU reserve) appeared to be reaching the end of its mine life in 2007 when it reported RAR sufficient to continue mining until 2014 (NEA–IAEA, 2008). Additional drilling has since identified 34,000 tU of high-grade RAR below the current openpit operation in this unconformity-type deposit (Mckay and Carson, 2010). A decline is being constructed to explore this new resource. In addition, a heap-leach facility is proposed for extracting a resource of 16,100 tU from low-grade RAR on stockpiles and from those remaining in the mine. The Ranger mine has experienced periodic delays in production for several weeks at a time during the rainy season when ore and treatment plants become inaccessible and facilities for impounding tailings have filled with water. Ranger operates within the Kakadu National Park, a United Nations Educational, Scientific and Cultural Organization World Heritage site. The operations are currently licensed through 2021, and any extension of operations beyond that time period would require new legislation. Production from Ranger has declined since at least 2007, and declines are likely to continue as the operation targets lower-grade ore through its use of surface heap-leaching and from resources lying deeper underground.

Jabiluka, a large unexploited 123,389 tU reserve, abuts the northern edge of the Ranger lease (IAEA, 2010). This property has been dormant since 2005, when its owner, Rio Tinto, reached an agreement with the Mirarr traditional aboriginal people that would require aboriginal approval prior to development. An exploration tunnel was backfilled with waste rock and unprocessed material, and the property was placed on standby. Development of Jabiluka will depend on the ability to overcome significant permitting impediments.

Another area of active mining and ongoing permitting and development are sandstone uranium deposits in the Fromme Embayment in South Australia. The Beverly ISL uranium mine (12,192 tU), owned and operated by Heathgate Resources Pty. Ltd. (Heathgate), is located in the Fromme Embayment, as are the following developing properties: Honeymoon (2,500 tU), owned by Uranium One, Oban (1,781 tU inferred), owned by Curnamona Energy Ltd., and Four Mile (23,462 tU), owned by Quasar Resources Pty. Ltd., and newly discovered Pepegoona (900 tU), also owned and operated by Heathgate (Mckay and Carson, 2010; Ux Consulting Company LLC, 2010). Protests against ISL mining as a procedure for recovering ores in Australia prompted the Australian government to investigate the impacts of leach mining (Commonwealth of Australia, 2010). The aquifer being mined is highly saline, and the area is remote from cities and farms so impacts on human health, on stock, or on agriculture from mining of uranium from this aquifer were determined to be minimal. Permitting and mining has proceeded fairly rapidly, and it is likely that uranium will continue to be produced from this region in response to ongoing exploration and development.

In Western Australia, BHP Billiton is conducting a feasibility study of its Yeelirrie deposit (44,077 tU inferred). This is a calcrete uranium deposit located 4 to 8 m below the surface that will be developed by surface operations in a progressive fashion with ongoing reclamation and remediation. Processing details have not yet been reported. This deposit is projected as producing between 2,000 and 2,500 tU per year over a mine life of 20 to 40 years. Two other calcrete uranium deposits in Western Australia are in the prefeasibility stage: the Wiluna project (9,385 tU) being explored by Toro Energy Ltd. and the Lake Maitland resource (10,000 tU), owned by Redport Ltd. (Mckay and Carson, 2010).

Brazil

Brazil's single operating uranium mine, the Lagoa Real–Caetite mine, a metasomatite deposit, produced 347 tU in 2009 from a remaining reserve of 12,700 tU (fig. 1–3; NEA–IAEA, 2010; Ux Consulting Company LLC, 2010)). An expansion is underway that will double production capacity by 2012. However, production is expected to fall significantly in the near future, due to licensing delays at a tailings dam (daSilva, 2010).

Figure 1–3. Active and developing uranium mines in Brazil. From Dahlkamp (2010).

A pilot plant to produce uranium from the Santa Quitéria/ Itataia phosphate deposit is being designed, constructed, and to be operational by 2012, with a capacity of 1,000 tU/yr. This resource is a substantial 67,240 tU (Ux Consulting Company LLC, 2010). As with all phosphate-uranium deposits, economic development depends on using an appropriate extraction technology, but none has been identified to date. Little exploration is being carried out in Brazil, although favorable geologic conditions exist (Dahlkamp, 2010). All uranium deposits in Brazil are controlled by the government owned Industrias Nucleares do Brazil.

Canada

Canada's three operating uranium mines produced 10,174 tU in 2009, providing 20 percent of the world's mined uranium (World Nuclear Association, 2011b). Production was from the rich unconformity-related uranium deposits in the Athabasca Basin of northern Saskatchewan. In 2009, the McArthur River mine produced 7400 tU, with a large remaining resource of 128,900 tU; McClean Lake produced 1,410 tU,

with a small remaining resource of 1,031 tU; and Rabbit Lake produced 1,400 tU, also with a relatively small remaining resource of 8,200 tU (fig. 1–4: (Ux Consulting Company LLC, 2010; World Nuclear Association, 2011b).

Deposits in the Athabasca basin that are in the development or feasibility stages include (1) Cameco's Cigar Lake mine, containing a significant reserve of 80,500 tU that is expected to produce in 2012 at an annual capacity of 6,294 tU; and (2) Areva's Midwest deposit, containing RAR of 16,340 tU with a planned capacity of 2,300 tU/yr (Calvert, 2010; Ux Consulting Company LLC, 2010). Cigar Lake, containing 80,500 tU in RAR was expected to be a significant uranium supplier by 2009, but a series of technical challenges that resulted in mine flooding and a total mine shutdown have delayed production until at least 2012. Water inflow is a significant problem in this wet northern environment, and the construction of freeze walls to control this water flow in active underground mines continues to be a technical and economic challenge. Midwest was originally planned as an underground mine, with startup in 2011, but an openpit option with no planned startup date is now reported (Calvert, 2010). Other advanced-stage projects in the Athabasca basin include Areva's Kiggavik (51,574 tU) and Cameco's Millennium (18,002 tU) deposits. Outside the Athabasca basin, Strateco Resources Inc. is exploring the 7,770-tU Matoush deposit in Quebec by surface drilling and construction of an underground exploration ramp scheduled for 2012. Aurora Energy Resources Inc., a subsidiary of Paladin, is working to develop two deposits in Labrador: the Jacques Lake (4,000 tU) and the Michelin (25,923 tU) deposits.

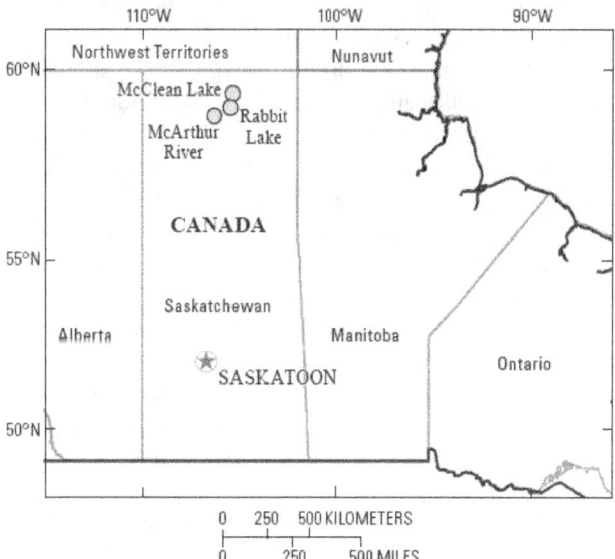

Figure 1–4. Active uranium mines in Canada. From WNA (2011).

Central African Republic

In 2010, a feasibility study of Areva's Bakouma deposit containing 9,885 tU was being conducted, with tentative plans to develop the mine by 2015 (fig. 1–5). In late 2011, however, Areva suspended the project for two years, stating that the mine was unprofitable at present uranium prices (Ngoupana and others, 2011). The Central African Republic retains a 10-percent interest in Bakouma (Areva NC, 2011).

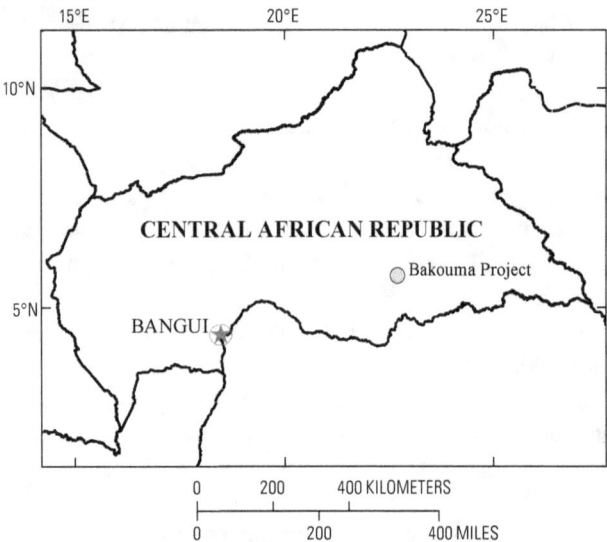

Figure 1–5. Location of the Bakouma deposit, Central African Republic. From Areva NC (2011).

China

All uranium mines in China are owned and operated by the China National Nuclear Corporation (CNNC). CNNC is a state-owned enterprise with hundreds of subsidiaries, and it fully controls the nation's nuclear fuel cycle of mining, conversion, enrichment, and fabrication. The "Red Books" for 2005, 2007, and 2009 (NEA–IAEA, 2006, 2008, 2010), Dahlkamp (2010) and Zhang, 2010 are this report's sources for resource and production data. Because of the limited reporting from China, there is some confusion about resource and development time lines.

Six production centers are reported for China: Fuzhou (Jiangxi Province), Chongyi (Giangxi Province), Yining (Xinjiang Province), Lantian (Shaanxi Province), Benxi/ Quinglong (Liaoning Province), and Shaoguan (Guandong Province) (fig. 1–6, table 1–1) (NEA–IAEA, 2010). Mines associated with the production centers are reported for the Yining ISL facility (Dep. 512); Lantian (Lantian deposit), Benxi (Benxi deposit), and Quinglong (Quinglong District). Uranium began to be produced from the Quinglong deposit in Liaoning Province through underground mining and surface

heap-leaching in 2007. Although the production capacity is listed as 100 tU/yr, Quinglong has not achieved full capacity because of lower than expected yields from heap-leaching. Production capacity and RAR are listed in table 3.

Mines reported as operating but with no associated production center include Xiangshan, Jiangxi Province; Xiazhuang, Guangdong Province ; Yili, Xinjiang province; Tengchong, Yunnan Province; Lianshanguan, Liaoning Province; Ziyauan, Guangxi Region; and Tengchong, Yunnan Province. Because Tengchong is listed as an ISL mine, production does not require a mill, and so no production center is associated with it (table 1–1).

Combining information reported in the 2009 "Red Book," in Dahlkamp (2010), and in the Uranium Suppliers Annual (Ux Consulting Company LLC, 2010), this report proposes the following associations of mine and mill, as based on the proximity of a production district where mines are listed as

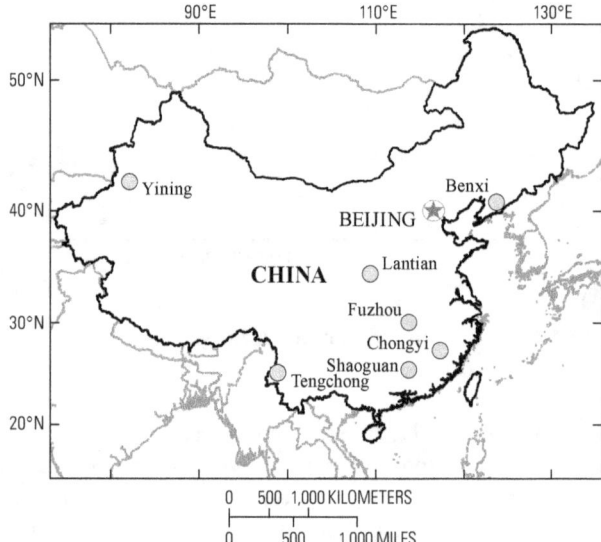

Figure 1–6. Major uranium production centers in China. From Dahlkamp (2010).

Table 1–1 Chinese uranium mines and associated production centers inferred for these mines.

Operating Uranium Mines and Production Centers in China		
Production center	**Reported associated mine**	**Inferred associated mine**
Fuzhou		Xiangshan
Chongyi		
Yining	Dep.512	Yili
Lantian	Lantian	
Benxi/Benxi	Benxi	Lianshanguan
Benxi/Quinglong	Quinglong	
Shaoguan		Xiazhuang
Operating mines with no associated production center		
Tengchong	(ISL—No milling required for production)	
Ziyauan		

operating: (1) the Fuzhou mill, which may be processing ore from deposits in the Xiangshan district; (2) the Yining mill, which may be processing ore from the Yili district; (3) the Shaooguan, milling ore from the Xiazhuang deposit; and (4) the Benxi, milling ore from the Lianshanguan deposit. This leaves the Ziyuan deposit for which no production center can be surmised.

Several deposits in China are described as being in some stage of feasibility. The 3,000-tU Sihongtan deposit near the Yining production area is an ISL deposit undergoing pilot testing. The 5,000-tU Dongsheng deposit in Inner Mongolia was determined not to be amenable to ISL mining, but feasibility studies of mining this resource using underground methods are being conducted. Feasibility studies are also reported at the 19,400-tU Erlian, the 17,000-tU Zaohuohao, and the 21,600-tU Erdos deposits in Inner Mongolia. The 5,000-tU Guyuan deposit in Hebei Province is in pilot testing or in construction. The Liaohe deposit of unknown size is also in the feasibility stage, according to the 2009 "Red Book". The 9,000-tU Turp-Hame deposit, possibly in the Turpan-Hami Basin that also hosts the Sihongtan deposit , is listed as an ISL project in the planning stages. The Shihongtan deposit in the Inrpan-Hami Basin in Xingiang Autonomour region is listed as in the feasibility stage in the IAEA Uranium Deposits database (IAEA, 2010), although the 2009 "Red Book" did not report this resource.

The 2009 "Red Book" reported that China has 171,400 tU in RAR from 13 deposits. Some of these deposits were described separately as operating mines, some as a depleted deposit, and some as being dormant; the status of several others is unknown. An additional "statistical" 1.2 to 1.7 million tonnes of "potential uranium resources are predicted" (NEA–IAEA, 2010). The status of many of these deposits is unknown; one is listed in the IAEA uranium deposits database as depleted (IAEA, 2010). Tables 2 and 3 list deposits believed to be operating or to be in some stage of feasibility. As a geologically diverse region, China might be expected to have a larger uranium resource than has been reported to date. The low numbers reported here for these resources may reflect lack of exploration, incomplete reporting, or the absence of a uranium-rich province within the country.

In addition to domestic resources, CNNC has an interest in RAR from the Azelik deposit in Niger; from the Gurvanbulag deposit in Mongolia; and from the Zhalpak, Irkol, and Semisbai deposits in Kazakhstan. The production and RAR for these deposits are described in this appendix under their respective countries.

Czech Republic

The Czech Republic has two operating uranium mines: the Straz mine in the Straz pod Ralskem district in northern Czech Republic, and the Rozna mine, part of the Rozna–Olsi

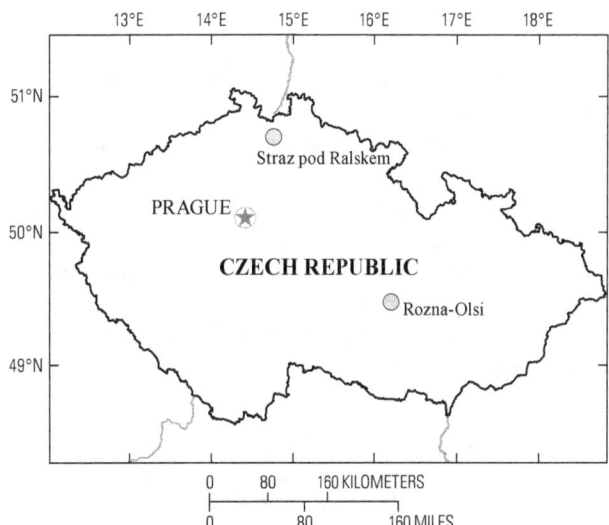

Figure 1–7. Uranium production centers in the Czech Republic. From OECD/NEA-IAEA (2010).

uranium field (fig. 1–7). Remaining RAR at Rozna are 680 tU, and at Straz they are 1,320 tU (NEA–IAEA, 2010). All uranium-related activities, including mining and environmental restoration, are carried out by the DIAMO state enterprise (DIAMO s.p.).

At Rozna, production using underground methods during 2009 was projected to be 255 tU. Mining is expected to continue through 2012 and beyond if it remains profitable. Production at Straz is a byproduct of remediation of this former ISL facility. In 2009, Straz produced 38 tU (100,000 pounds. U_3O_8) and production is decreasing as environmental cleanup at the mine site continues while uranium concentrations in leachate solutions decrease. At Straz more than 4.5 million tonnes of sulfuric, nitric, hydrochloric, and hydrofluoric acid were used to mine this deposit during the 30-year period 1967–1996. Cross-contamination between aquifers allowed mining solutions to migrate into the drinking-water aquifer that the local community relied on. In 2009, the Czech Ministry of Economics proposed that mining be resumed at Straz either by underground or ISL techniques.

Government policy set in 1980 in the Czech Republic was to eventually close all of the nation's uranium mines, and for this reason no new mines are anticipated to open in the near future. However, uranium resources are believed to exist in several regions. Of the 23 identified Czech Republic deposits, 20 have been mined, Rozna is currently being mined, and the remaining two identified deposits, Osecna-Kotel and Brzkov, remain unexploited. In 2008 ,the Australian mining company, Uran Ltd. (Uran), applied for a lease to initiate exploration around the 7,500-tU Brzkov deposit in the central Czech Republic, but the Environment Ministry did not grant the license. Uran also attempted to recover uranium from surface-dump rock at the Pribram mine, but whether this venture was successful is unknown.

Denmark (Greenland)

The multielement Kvanefjeld deposit (Greenland Minerals and Energy Ltd.) contains an identified 85,614 tU, as well as rare-earth elements and zinc in mineable quantities (fig. 1–8: (NEA–IAEA, 2010). Several other exploration targets and prospective terrane exist throughout the country. Currently, Denmark has banned uranium exploration and exploitation in Greenland. Although a feasibility study has been performed on the Kvanefjeld deposit, the in-place ban on mining prevents this report from including Kvanefjeld among world uranium resources.

Finland

Although there is considerable exploration for uranium in Finland, there is no current production. The Talvivaara polymetallic black shale deposit, in the Kainuu Province of eastern Finland is mined for nickel and zinc that contain trace uranium (0.001 to 0.004 percent U) (fig. 1–9). The mine operator, Talvivaara Mining Company Plc, is constructing a uranium extraction circuit at the mine that is expected to be complete in 2012. Cameco has signed an agreement to help finance production and to buy this uranium, which is estimated to be produced at a rate of 350 tU/yr.

Figure 1–8. Location and geology of the Kvanefjeld deposit, Greenland. From Greenland Minerals and Energy Ltd. (2011).

Figure 1–9. Location of the Talvivaara uranium deposit in Finland. From Talvivaara Mining Company Plc., (2011).

India

All uranium mines and production are owned and controlled by Uranium Corporation of India, the India State mining company. Six operating mines are reported: Narwapahar (11,500 tU); Jaduguda (8,400 tU); Turamdih and Banduhurang (3,750 tU); Bagjata (2,106 tU); and Bhatin (2,200 tU) mines (Chaki, 2010; IAEA, 2010) (fig. 1–10). Since India does not regularly report RAR remaining in these mines, the available resource figures probably do not account for depletion by mining and should be considered high. Three mines are in development: the Mohuldih (unknown resource), Lambapur-Peddagattu (unknown resource) and the Tummalapalle (12,555 tU) deposits. The RAR estimated to be remaining in operating and developing mines in India total 40,511 tU. The 2009 "Red Book" reports a slightly larger number (55,200 tU) for the country, but it does not allocate quantities to individual mines. Other mine development reported for India is still in the planning stages, and is not included in this summary. Uranium has been recovered from the Rakha and Mosaboni mines as a product of copper mining, but it is not currently being produced from these mines (Ux Consulting Company LLC, 2010).

Because India has historically been unable to import uranium, the Uranium Corporation of India has explored and continues to aggressively explore prospective terrane in order to provide fuel to its 19 operating powerplants. India is also actively exploring the viability of thorium-fueled rather than uranium-fueled reactors because the country has a significant thorium resource. It does not seem likely that India will supply uranium to the world market in the immediate future.

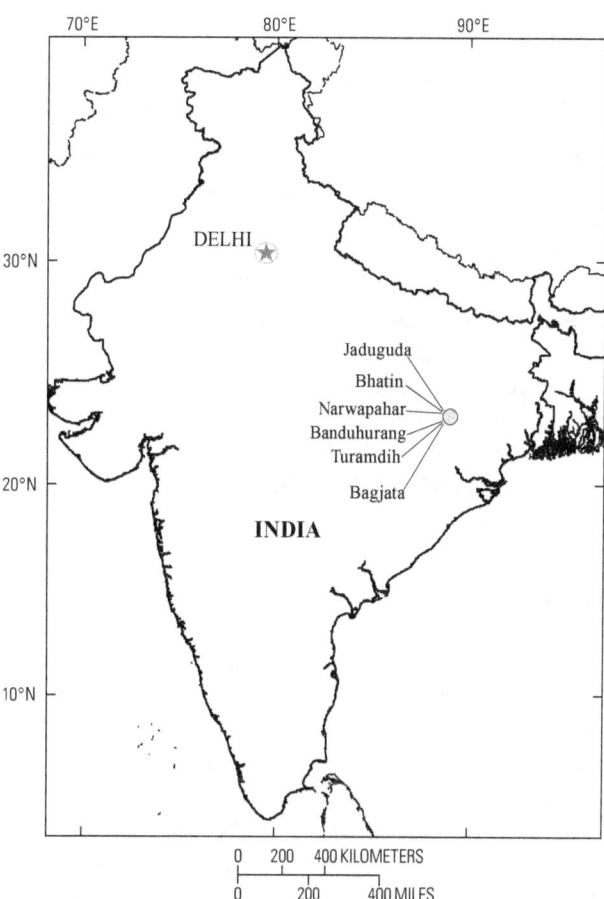

Figure 1–10. Location of current production centers and prospective areas of India. From Chaki (2010).

Iran

Exploration and development in Iran is performed by the government agency, Atomic Energy Organization of Iran, which reports one existing and one committed production center (NEA–IAEA, 2010). Production capacity is 21 tU from a 100-tU reserve at the Gachin mine, although since at least 2007 this reserve has been reported with no measurable reduction from having been mined (fig. 1–11). The small size and low grade of Gachin ore would probably make this deposit subeconomic without Iran's strong desire for a domestic source for uranium. Development of an underground (90 percent) and openpit (10 percent) mine to exploit the 900-tU Saghand deposit is underway, with production at a rate of 50 tU per year planned to commence in 2012. The 2009 "Red Book" reports a strong exploration program, utilizing modern techniques and targeting several different geologic provinces and deposit types. It is not anticipated than Iran will contribute uranium to the world market in the near future because mined uranium will be used domestically.

Figure 1–11. Location of the Gachin mine and the Saghand uranium deposit, Iran. From Dahlkamp (2010).

Jordan

Jordan does not currently produce uranium, and has no deposits that can be classified as imminent producers. However, a number of uranium companies, including France's Areva and Australia's Rio Tinto, are actively exploring in this country. As well, the Jordan Atomic Energy Commission's commercial arm Jordan Energy Resources Inc. (JERI) is conducting regional exploration to encourage commercial development of uranium deposits in the country. A total resource of between 59,360 and 165,470 tU is identified in phosphate deposits in Jordan. In addition, surficial deposits ranging from 50,000 to 70,000 tU in are currently being explored in central Jordan south of Amman by JERI (NEA–IAEA, 2008; IAEA, 2010; Ux Consulting Company LLC, 2010).

Kazakhstan

The national mining company, KazAtomProm, is responsible for all uranium mining activities in Kazakhstan, the world's largest uranium producer. KazAtomProm has formed joint-venture partnerships with the following mining companies to develop properties: Areva, Cameco, Uranium One, Sumitomo Corporation, Energy Asia Ltd., and CNNC.

Fifteen operating mines and two pilot projects were reported for Kazakhstan in 2010 (tables 3, 4). The mines are located in three mining districts: (1) the two pilot projects in Kokshetau Region (Tselinny, and Semisbai); (2) the 12 mines in the Chu-Sarysu district (Akdala, Centralnoye, Chiili, Stepnoye, Inkai, S. Inkai, Muyunkum/Tortkuduk, Zarechnoye, Central Mynkuduk, W. Cynkuduk, Budenovskoye 2, and Budenovskoye 1,3,4); and (3) the three mines in the Syr-Darya district (Kharasan 1, Kharasan 2, Irkol) (NEA–IAEA, 2010) (fig. 1–12). ISL mining using an aggressive sulfuric acid lixiviant is the most common mining method in the primarily sandstone-hosted roll-front deposits. Production costs are low, permitting and restoration hurdles are not high, and the country is aggressively promoting the development of deposits, the result being that uranium production is expanding rapidly. With a large national resource of 459,677 tU, Kazakhstan is expected to continue as a top producer for the immediate future.

Malawi

Malawi reports one operating uranium mine, the Kayelekera deposit that is being mined using openpit methods by Paladin (fig. 1–13). Paladin is producing 1,270 tU/yr from the 11,265-tU deposit, which it expects to deplete by 2020 (Paladin Energy Ltd., 2011).

Figure 1–12.   Location of major uranium mining districts and operating mines of Kazakhstan. From Dahlkamp (2010).

Figure 1–13.   Location of the Kayelekera Mine in Malawi. From Paladin Energy, Ltd. (2011).

Mongolia

One advanced-staged project is reported for Mongolia, the Dornod deposit containing 24,780 tU, which is expected to be mined by Khan Resources Inc. at a rate of 1,159 tU/yr during the 10 years beginning about 2015 (fig. 1–14) (NEA–IAEA, 2010). The deposit was partly mined by the Russian Federations' Priargunsky Mining and Chemical Enterprise during 1988–1995, when infrastructure, including a rail line over which ore was shipped 500 km to Krasnokamensk in Siberia, was developed at the mine site. The Nuclear Agency of Mongolia delayed development of the Dornod deposit by invalidating Khan's mining license in 2009 (Khan Resources Inc., 2011). Litigation ensued, and has not been fully resolved. Uranium may be produced from Mongolia from this deposit in the near future if title issues for Dornod are successfully resolved so that Khan can begin raising sufficient financing for the project.

Four mining companies are involved in uranium exploration throughout Mongolia: Cameco, East Asia Minerals Corp., Areva, and Solomon Resources Ltd. No other projects are far enough advanced to anticipate production in the near future.

Namibia

Following the passage of the Minerals Act of 1992 that established uranium as a strategic mineral, the government of Namibia began working to develop mining guidelines for studying the cumulative societal and environmental impacts of mining in the west-central Erongo region, the major uranium province of Namibia. Overwhelmed with applications while working to develop a national policy for uranium mining, Namibia in 2007 called a moratorium on granting new exploration licenses.

Two mines are producing uranium in Namibia; Rio Tinto's Rossing Mine (50,657 tU resource) with a production capacity of 3,817 tU/yr; and Paladin's Langer Heinrich mine, with a production capacity of 1,425 tU/yr from a 60,830 tU reserve (fig. 1–15) (NEA–IAEA, 2010; Ux Consulting Company LLC, 2010). Rossing has been in operation since 1976, and has a life-of-mine operating plan that details production through 2023. If mine-site exploration results are encouraging, mine life may be extended. Rossing is testing enhanced heap-leach techniques that may increase production. At Langer Heinrich, Paladin is capitalizing on recent increases in RAR and is planning a plant expansion that would increase production to 3,500 tU/yr.

Areva is developing the Trekkopje deposit, with tentative plans to begin production from a 42,243-tU reserve at an initial rate of 1,600 tU/yr by 2013, eventually ramping up to 3,500 tU/yr. A mine license was granted in 2008, and pilot testing is underway. The mine life of Trekkopje is expected to be 12 years, until 2023. Also close to production is the 23,269-tU Valencia deposit, which Forsys Metals Corp. is developing with plans to produce 1,000 tU/yr by 2013. The

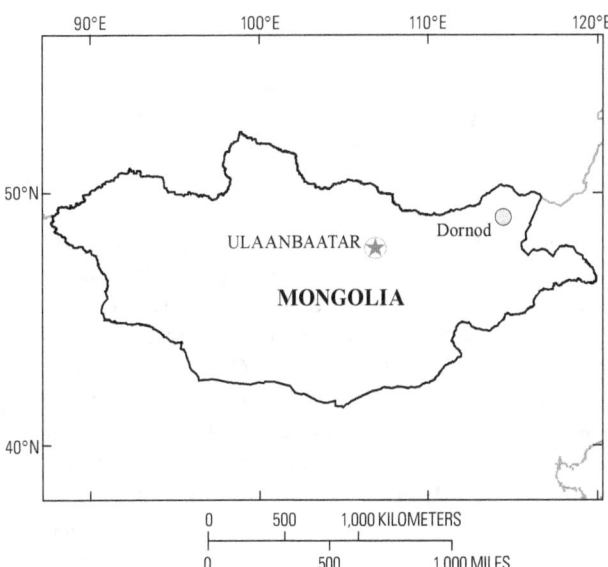

Figure 1–14. Location of the Dornod Uranium Project, Mongolia. From Khan Resources Inc. (2011).

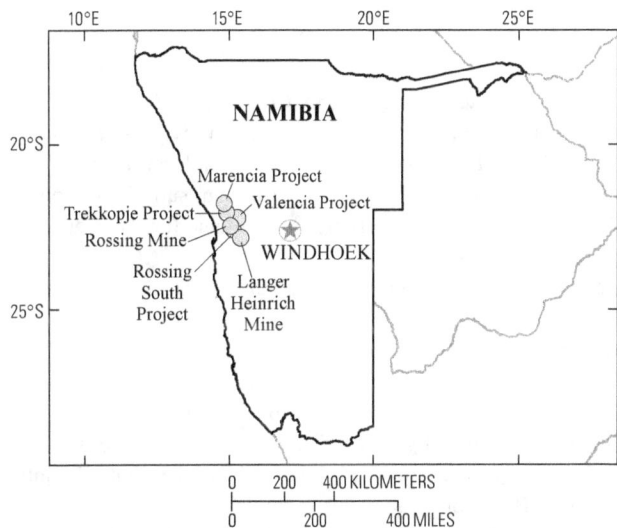

Figure 1–15. Approximate location of operating uranium mines and advanced-stage projects in Namibia. From OECD/NEA-IAEA (2010) and from Marencia Energy Limited (2011).

Marencia uranium deposit (62,856 tU) is being developed to produce at a rate of 1,000 tU/yr by Marencia Energy Ltd. by 2014, with financing through China's Hanlong Energy Ltd. (Ux Consulting Company LLC, 2010). The Rossing South deposit, currently owned by Extract Resources Inc. and part of the larger Husab deposit, contains an estimated 98,846 tU. The Husab deposit is a significant future producer, but was not included in this analysis because it is still in the exploration stage (Ux Consulting Company LLC, 2010). Other projects in Namibia are progressing but are still many years away from production (Itamba, 2011).

Niger

Two producing centers, both operated by Areva, are reported for Niger. The 24,670-tU Akouta mine with a production capacity of 2,000 tU/yr, and the 23,171-tU Arlit mine with a production capacity of 2,000 tU/yr (fig. 1–16: (NEA–IAEA, 2010; Ux Consulting Company LLC, 2010).

The Imouraren and Azelik advanced-stage projects are projected to start up by 2013. Areva Company's Imouraren mine has a reserve of 183,520 tU, the second largest reserve in the world, and its production capacity is proposed as ranging from 2,000 to 7,000 tU/yr (table 3) (NEA–IAEA, 2010; Ux Consulting Company LLC, 2010). The Imouraren mine is expected to produce uranium for 35 years, through 2048. The Azelik deposit, with 10,800 tU, is jointly being developed to its proposed capacity of 1000 tU/yr by the Nigerian government and a private Nigerian group, Trendfield Holdings SA (NEA–IAEA, 2010; Ux Consulting Company LLC, 2010). Other mining companies are actively exploring tenements in Niger, but none has reported reaching the feasibility or development stage.

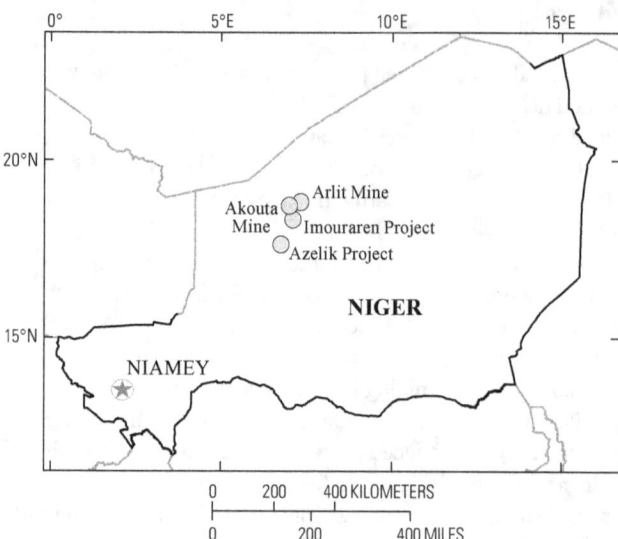

Figure 1–16. Location of uranium mines and advanced stage projects in Niger. From OECD/NEA-IAEA (2010) and Areva NC (2011).

Pakistan

The Pakistan Atomic Energy Commission (PAEC) controls all uranium mining within the country. Pakistan is nearing the end of a 5-year (2006–2011) exploration program designed to identify domestic uranium supplies nationwide. No recent country reports are available for Pakistan, so estimates of national RAR are based on the IAEA world distribution of uranium deposits database (UDEPO) and on Ux Consulting Company's Uranium Suppliers Annual (IAEA, 2010; Ux Consulting Company LLC, 2010).

Three mines are reported to be operating in Pakistan: the Quabul Khel/Issa Khel, the Tumman Leghari, and the Dera Ghazi Khan mines (fig. 1–17). The Tumman Leghari mine could not be located, but is reported to be in the South Punjab Province. RAR for these deposits are unknown. There is some confusion in the literature about the location of the Quabul Khel and Issa Khel mine/production center (Dahlkamp, 2010; Ux Consulting Company LLC, 2010). It is likely mining is being carried out at Quabul Khel, with the ore being processed at Issa Khel, so the location of both are show on figure 1–17. The cumulative nominal production capacity of Pakistani mines is estimated to be 84 tU. One deposit, the Shanawah ISL mine, is in development and is expected to be producing at a rate of 50 tU by 2014 from a resource of 2,578 tU (IAEA, 2010). PAEC claims to have located one thousand favorable uranium areas in the country.

Romania

The Romanian government's Uranium National Company (UNC) owns and manages exploration, mining, and production of uranium resources. Uranium production in Romania has

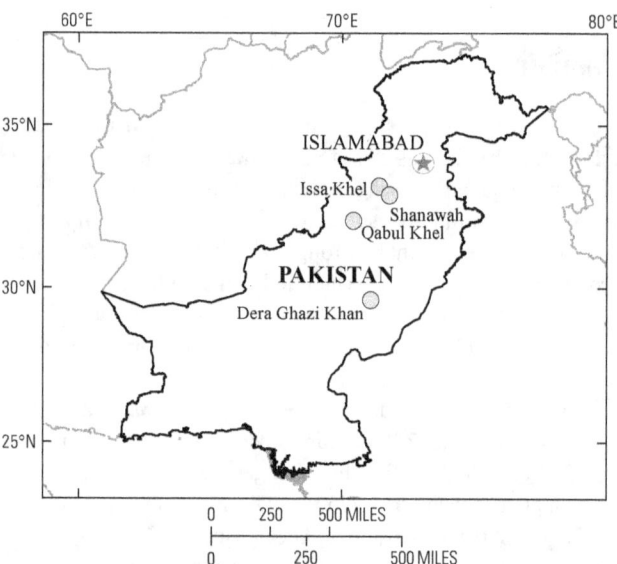

Figure 1–17. Location of uranium mines, production centers and advanced properties in Pakistan. From Dahlkamp (2010).

decreased significantly during the past decade, due to depleted resources and closing of unprofitable production facilities. The "Energy Strategy of Romania," a report on domestic reserves spanning the years 2007–2020, states that RAR are sufficient to supply domestic nuclear energy requirements for the 10 years through 2020 (Ux Consulting Company LLC, 2010). In 2010, the Romanian government provided UNC with subsidies for production of 45 tU (Ux Consulting Company LLC, 2010).

Romania's last submission to the NEA "Red Book" was in 2002. In 2009, the IAEA Secretariat estimated Romania's reasonably assured resources (RAR) at 3,100 tU, and its inferred resources at 3,600 tU (NEA–IAEA, 2010). Other

estimates report a total reserve of 8,769 for mines in the
Banat Region, Bihor (Apuseni Mountains), and Crucea
(Eastern Carpathians) regions (fig. 1–18) (Ux Consulting
Company LLC, 2010). Past and present production of
38 tU/yr is expected to continue through 2019 (Ux Consulting
Company LLC, 2010). Reports in 2009 indicate that pro-
duction is coming from the Crucea Mine in the Eastern
Carpathians (NEA–IAEA, 2000).

Russia

The Russian Federation is among the top five nations
producing uranium, with a total cumulative production
of 139,735 tU as of 2008. Russia reported RAR of over
181,000 tU, a total that could potentially double after
feasibility studies are completed for inferred resources that
exceed 300,000 tU (NEA–IAEA, 2010). Most of Russia's
uranium is produced from the Priargunsky mining complex, the world's largest uranium-producing center, which
has produced a total of 130,000 tU as of 2008 (fig. 1–19)
(NEA–IAEA, 2010).

Six Russian uranium deposits are significant. (1) The
Streltsovsk district Mining and Chemical Works (PCMCW),
(also referred to as "the Priargunsky works") has volcanic
caldera deposits where mining is ongoing and for which
NEA reports RAR of 102,600 tU, and IR of 26,930 tU, and
for which Ux Consulting Company LLC (UXC) reports
118,341 tU. (2) The Dalur production center (Dolmatovskoye
ISL mine) is producing uranium; the 2009 "Red Book" reports
RAR of 10,970 tU. (3) The Khiagda deposit, which is amenable to ISL mining, has RAR reported as 26,805 tU and has not
yet been mined. (4) Vein deposits of the Gornoe deposit have
RAR reported as 7,918 tU, and with production anticipated by
2014. (5) Vein deposits of the Olovskaya deposit are reported
as 12,200 tU, with production anticipated by 2014. (6) The
metasomatic Elkon district contains large inferred resources
(319,00 tU), with RAR reported as 71,300 tU, with production anticipated by 2015. (NEA–IAEA, 2010; Ux Consulting
Company LLC, 2010). Both the Streltsovsk (PCMCW) and
Dalur mines are producing uranium, with production anticipated from Olovskaya and Gornoe by 2014, and from Elkon
by 2015 (NEA–IAEA, 2010).

The Russian government plans to continue to expand
uranium resources, production, and all other steps in the fuel
fabrication process to meet the growing needs not only of
the Russian nuclear industry but of the global fuel market.
In primary supply, their goal is to increase uranium production capacity from 3,521 tU in 2008 (NEA–IAEA, 2010) to
12,000 tU by 2026 (Ux Consulting Company LLC, 2010).

South Africa

The South African government plans to increase domestic nuclear power generation and to become self-sufficient
in all steps of the nuclear fuel cycle (NECSA, 2010). The

Figure 1–18. Major uranium mining regions of Romania.

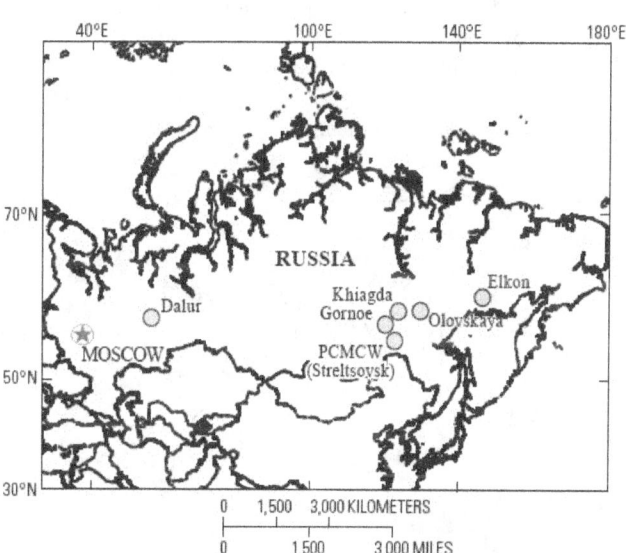

Figure 1–19. Location of major Russian uranium mining and
processing districts. From OECD/NEA-IAEA (2010).

government reports an increase in the uranium resource base
through renewed brownfield exploration efforts during the
five years 2006–2011, and production capacity is expected to
double by 2013 (NECSA, 2010).

In South Africa, uranium has been primarily a byproduct
of gold mining. Since production began in 1951, most
uranium was from underground mines within quartz-pebble
conglomerates in the Archean Witwatersrand Basin in the
northwest region of the country. Potential for uranium producers expanded quickly in 2007, when brownfield exploration,
stimulated by the "uranium beneficiation" program of the
South African Government, increased the number of operating
and planned uranium-producing mines.

The two main producers of uranium, the Vaal River and Ezulwini mines, are located within the Witwatersrand Basin, and their combined RAR are reported to contain 17,076 tU (fig. 1–20) (Ux Consulting Company LLC, 2010). The Ezulwini underground mine, about 40 km southwest of Johannesburg, is owned and operated by the First Uranium Corporation, which estimates more than 4,600 tU in proven and probable reserves, and in measured and indicated resources from underground deposits and from mine-waste tailings (NEA–IAEA, 2010). This report uses a more conservative estimate of measured and indicated resources for Ezulwini: 2,730 tU (Ux Consulting Company LLC, 2010). Production is anticipated to be 500 tU/yr by 2012 (NECSA, 2010), the increase in production being attributed to expanding underground development at the Ezulwini mine and also to expanding plant capacity of the Mine Waste Solutions (MWS) tailings recovery facility. The Ezulwini mine is expected to produce through 2030, and the MWS operation is expected to begin producing in 2012, continuing through 2026 (Ux Consulting Company LLC, 2010).

The second main producing mine, the Vaal River mine complex near Klerksdorp, is owned by AngloGold Ashanti Ltd. (fig. 1–20). At Vaal River, three mines are producing uranium as a byproduct to gold production: the Great Noligwa, the Kopanang, and the Moab Khotsong mines. They report RAR of 14,346 tU (NECSA, 2010; Ux Consulting Company LLC, 2010). Anglogold Ashanti Ltd. is the largest producer

of both gold and uranium-byproduct in the country, reporting RAR of 70,500 tU (IAEA, 2010). The company is expanding the reserve base through exploring extensions of known ore bodies, and it is doubling capacity of its uranium plant by refurbishing and rehabilitating existing facilities. Production exceeded 600 tU in 2009 and is expected to reach more than 1,200 tU by 2012, leveling out at 1,100 tU in 2014, and maintaining this level through 2025.

The Dominion Reef deposit is currently on standby. This property was acquired by Shiva Uranium Pty. LLC in 2010, with plans to rehabilitate the production plant and to resume mining in 2012. Reserve estimates total 55,753 tU, at an average grade of 0.062 percent, with expected production of 425 tU/yr for 2012. Plans are to increase production to between 850 and 1,460 tU a year by 2013, and to maintain this level through 2025 (NECSA, 2010). After Shiva's purchase of Dominion, however, significant progress toward production has not been reported for this mine.

Two mines will potentially come online in 2012 and 2017: the Cooke and the Ryst Kuil deposits. Using conventional underground mining methods and processing of tailings, the Rand Uranium Company is developing the Cooke property, which is outside Johannesburg in the Randfontein region within the Elsburg and VCR Reef deposits. The company was formed in 2008 to expand historic gold-producing mines into large-scale uranium and gold operations, capitalizing on surface resources contained within the Cook Tailings Dam. During the three years 2012–2015, the company is focusing on developing a uranium processing plant, expanding underground mining, and recovering uranium from tailings. Production of this Cooke resource of 9,464 tU is scheduled to commence in 2012 at a rate of 425 tU/yr. Underground mine life is expected to last at least 10 years through 2022, and the tailings feed to last for 17 years, to 2030.

The Ryst Kuil sandstone deposit (Areva joint venture with AngloGold Ashanti Ltd.) in the Karoo region has historic RAR of 7,731 tU (Ux Consulting Company LLC, 2010). Current exploration is underway, with the goals of updating estimates of reserves and of resources, starting commercial production in 2017–2018 at a rate of 1,136 tU/yr, and producing molybdenum as a significant byproduct (NEA–IAEA, 2010; Ux Consulting Company LLC, 2010).

Other less certain plans for developing South African uranium resource prospects involve properties owned by Mintails Ltd., Witwatersrand Consolidated Gold Resources Ltd., Niger Uranium SA, and Harmony Gold Mining Company Ltd., with reported measured and indicated resources totaling 35,196 tU (table 3).

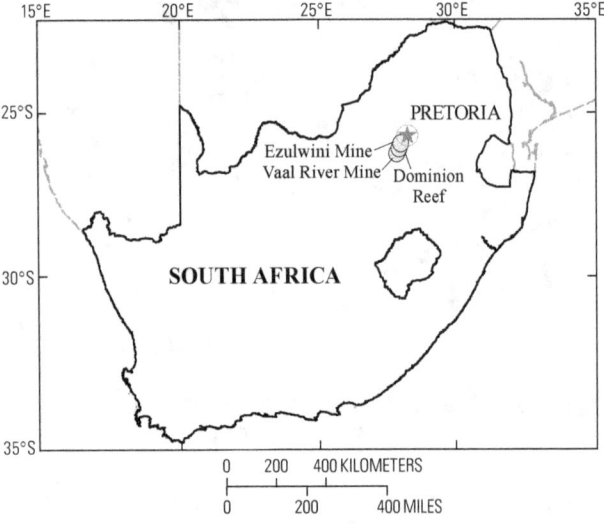

Figure 1–20.   Locations of the two operating mines (Ezulwini and Vaal River) and of one mine on standby (Dominion) in South Africa. From OECD/NEA-IAEA (2010).

Spain

Spain reports 2,500 tU in the cost category of <USD 80kg/U, and it reports 2,400 tU in the cost category of <USD 130/kgU (NEA–IAEA, 2010). A past producer, there is currently no uranium production from Spain. Berkeley Resources Ltd. (Berkeley) is actively exploring in the country in the Salamanca and Caceres areas (fig. 1–21). Berkeley also holds mineral tenements in the Calaf region where uranium is found in lignite seams. In 2009 Berkeley reached a collaborative agreement, along with ENUSA Industrias Avanzadas SA with Spain's Council of Ministers to complete a feasibility study at the Salamanca I deposit (NEA–IAEA, 2010) Salamanca I is anticipated to produce 769 tU/yr from a 30,926-tU deposit, beginning in 2014 (Ux Consulting Company LLC, 2010).

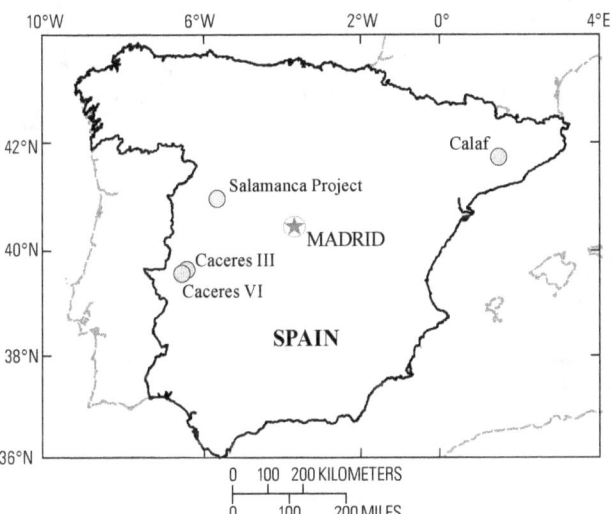

Figure 1–21. Location of active uranium exploration projects in Spain. From Berkeley Resources Ltd. (2011).

Ukraine

The Ukraine government is in the process of increasing domestic nuclear capacity and uranium production, with plans for doubling 2010 domestic production by 2013. Although domestic uranium production supplied 30 percent of Ukraine's nuclear energy requirements in 2008, increased domestic uranium production is expected to supply 100 percent of the country's nuclear energy requirements by 2012 (NEA–IAEA, 2010). All aspects of the uranium and nuclear industry are state-owned and controlled, operating under the Vostochny Integrated Mining and Concentrating Plant (VostGOK).

In 2008, Ukraine reported 161,601 tU in the RAR category. To date (2011), most of Ukraine's uranium (>100,000 tU) has been extracted from metasomatite deposits within the Kirovograd block of the Ukrainian Shield, using conventional underground mining methods and underground block leaching techniques. Remaining RAR deposits are estimated at about 142,000 tU, at grades of 0.1–0.2 percent U (NEA–IAEA, 2010; Ux Consulting Company LLC, 2010). The remaining RAR—6,900 tU in sandstone deposits within the sedimentary cover of the Ukrainian Shield, at grades of 0.01–0.06 percent U—are amenable to ISL mining.

Production centers include (1) the Hydrometallurgical Plant (HMP) in Zheltiye Vody plant, which has been operating since 1958 and currently produces 1000 tU/yr from the Michurinskoye, Central and Vatutinskoye deposits through the processes of acid leach, solvent extraction, and ion exchange (NEA–IAEA, 2010); (2) the new processing facility Novokonstantinovskoye HMP in the Kirovograd District, which is expected to reach a capacity of 2,500 tU year by 2015 (NEA–IAEA, 2010); and (3) the Safonovskoye ISL plant in the Kazanofsky District, with planned production of 150 tU/yr in 2012 (Ux Consulting Company LLC, 2010) (fig. 1–22).

Figure 1–22. Uranium production centers in Ukraine. From Bakarzhiyev, (2011).

United States

The following description of U.S. uranium resources includes only the publicly available estimates of resources associated with operating or with developing mines; it does not include United States RAR that are included in other estimates by the EIA uranium reserve , because the EIA considers this information to be proprietary and therefore protected.

Mines in the United States produced 3.7 million pounds. U_3O_8 (1,577 tU) in 2009 and 4.2 million pounds U_3O_8 (1,615 tU) in 2010 (Energy Information Administration, 2011). During the fourth quarter of 2011, five ISL operations, with a combined reserve of 14,737 tU, were in production: the Alta Mesa and the La Palangana mines in Texas, the Crow Butte operation in Nebraska, and the Smith Ranch–Highland mine, and the Willow Creek mine in Wyoming. Several small conventional mines on the Colorado Plateau operated intermittently during this time, with all ore processed at the Denison Mines Corp. (Denison) White Mesa Mill in Blanding, Utah (purchased by Energy Fuels Inc. in 2012).

In-situ leaching operations produce most of the uranium concentrate in the United States, and most developing mines in the country use this process. Currently operating

ISL plants report more than 7,000 tU of demonstrated economic resources, and more than 61,000 tU are associated with developing ISL operations. Uranium-bearing sedimentary deposits amenable to ISL are in abundance in the Gulf Coast Province in Texas, in Wyoming basins, in the Crawford Basin of Nebraska (Black Hills-Northern Great Plains Province), and in the Grants district of New Mexico (fig. 1–23).

The United States has significant uranium deposits in areas where geologic factors require mining of uranium ore by conventional underground or openpit mining techniques, such as the high-grade breccia pipes of the Arizona Strip and the uranium-bearing sandstones of the Uravan Mineral Belt. Development of underground and openpit uranium mining is limited by higher costs for extraction, transportation costs, and limited milling capacity. In December 2010, operating underground mines within the Colorado Plateau, the Arizona Strip, and the Uravan Mineral Belt contained more than 3,000 tU in economic resources. Developing underground and openpit mines report more than 12,000 tU in demonstrated economic resources. Ore from these mines is currently processed at the White Mesa Mill, which has an operating capacity of 2,000 tU/yr. In 2011, the White Mesa Mill was supplied with ore mined from underground mines from the Arizona Strip

United States Uranium Provinces, Districts, and Important Deposits

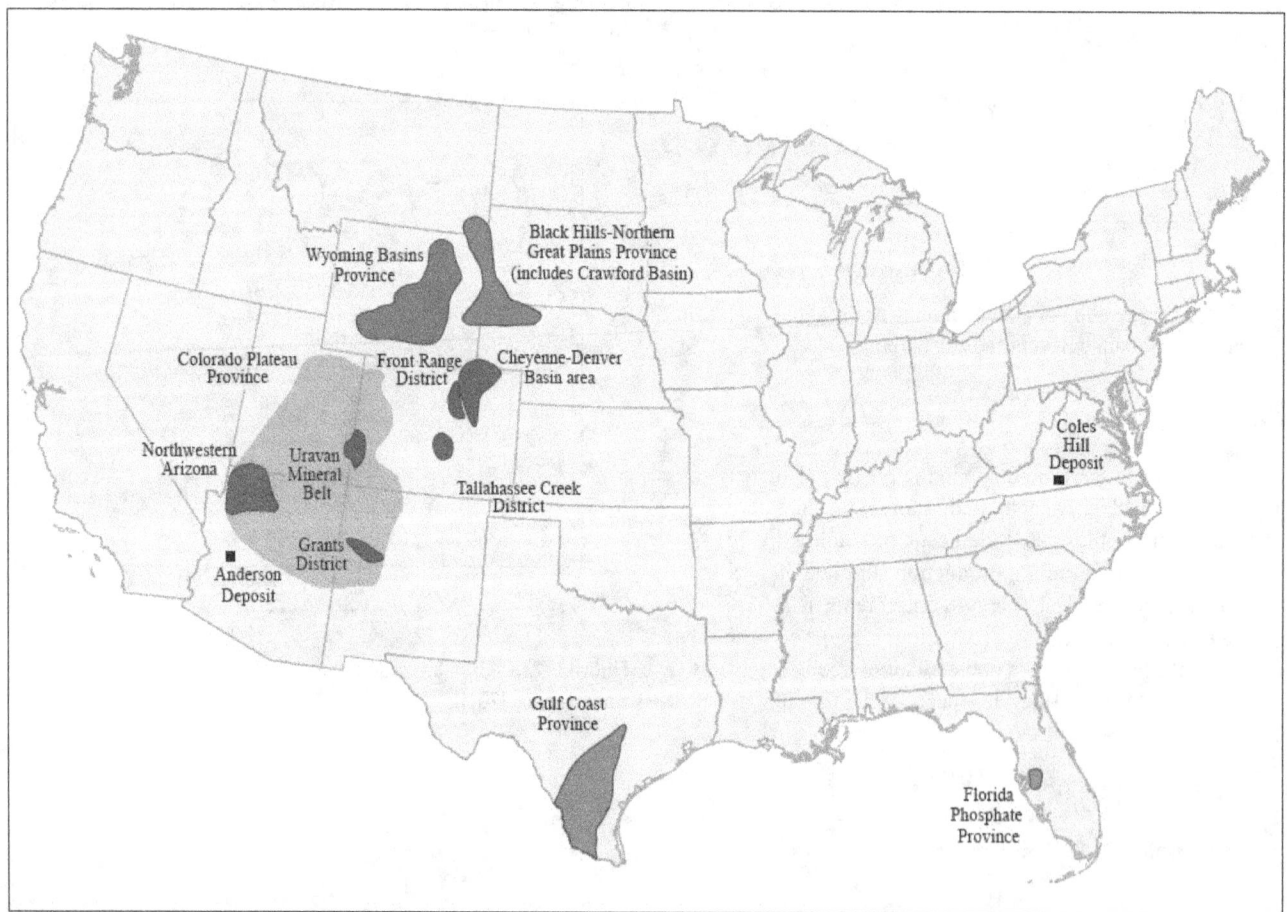

Figure 1–23.   Uranium province map of the United States.

(Arizona One mine) and from the Colorado Plateau District (Pandora and Daneros mines). Three other existing mills with a total capacity of 4,150 short tons of ore per day were on standby: the Canon City Mill in Colorado, the Sweetwater Uranium Project Mill in Wyoming, and the Shootaring Canyon Uranium Mill in Utah. One planned mill in western Colorado, Energy Fuel's Piñon Ridge, is almost fully permitted (Energy Fuels Inc., 2012).

The development timeline for mines in the United States is long, and the major hurdle to production continues to be regulatory uncertainty. The Coles Hill deposit in Virginia (Virginia Uranium Inc.; 27,656 tU) cannot currently be developed due to the State's moratorium on uranium mining and is therefore not included in this review of world supply. Development is proceeding at a much more rapid pace in States considered friendly to mining, such as Wyoming, whereas development is at a virtual standstill in States with active groups opposed to uranium mining, such as New Mexico.

Uzbekistan

Uzbekistan is one of the world's top 10 uranium producers. The country has no national uranium requirements and exports all the uranium it produces. The Navoi Mining and Metallurgical Complex (Navoi MMC), owned by the Uzbekistan government, controls all aspects of uranium resource exploration and production.

Uzbekistan's uranium resources are in sandstone and black shale deposits, with production only coming from the sandstone deposits. Cenozoic and Mesozoic sedimentary basins host the uranium-bearing sandstones in a 125-km-wide belt that extends more than 400 km from Uchkuduk in the northwest to Nurabad in the southeast. All current and planned production is by *in-situ* leaching from sandstone deposits. The black shale deposits are breccia complexes hosted by deformed Precambrian-Paleozoic carbonaceous and siliceous schists. Mineralization within the black shales includes uranium–vanadium–phosphate ores and could be mined by open-pit and by heap-leaching, but there are no immediate plans (as of 2005) to mine these deposits (NEA–IAEA, 2006).

Uranium is produced by Navoi MMC from three production centers that have all been operating since 1960s: (1) the Northern Mining Division (Uchkuduk District), (2) the Southern Mining Division (Zafarabad District), and (3) Mining Division #5 (Nurabad District) (Dahlkamp, 2010) . Annual production all from ISL mines and was 2,300 tU in 2005, with plans to increase to 3,000 tU/yr through 2040 (fig. 1–24).

Uzbekistan has not reported uranium resource information to the NEA since 2005. The 2009 "Red Book" reports resource estimates based on the IAEA Secretariat's adjustments of the 2005 data reduced by past production. For 2010 NEA estimates RAR of 76,000 tU and IR of 38,600 tU (NEA–IAEA, 2010). The Ux Consulting Company LLC reports RAR of 108,441 tU, which is fairly consistent with NEA's reporting (Ux Consulting Company LLC, 2010).

Figure 1–24. Active uranium mining districts in Uzbekistan. From NEA–IAEA (2006).

Other Countries

Nineteen countries, in addition to these 24 discussed above, are listed in the 2009 "Red Book" as containing RAR but as currently not producing uranium (table 6) (NEA–IAEA, 2010). For example, all the following countries contain resources but the resources are in the highest cost category of USD 260/kgU: France (9,000 tU), Mexico (1,300 tU), Tanzania (8,900 tU), Slovakia (5,100 tU), Somalia (5,000 tU), and Vietnam (1,000 tU). These economic barriers may well account for the lack of production (NEA–IAEA, 2010).

Algeria reports 9,500 tU in the top two cost categories. Significant exploration has been carried out in this country, but no uranium production is reported (NEA–IAEA, 2008). Chile last reported RAR of 1,034 tU in an undesignated cost category (NEA–IAEA, 2008). These resources are in low-grade surficial deposits (~ 0.02 percent U), in metasomatic deposits (0.03–0.20 percent U), and in one volcanic deposit, El Laco (0.01–0.18 percent U). In 2009, the IAEA Secretariat revised this estimate, adjusted for recovery factors, to 800 tU in the highest cost category (NEA–IAEA, 2010). Although no production has been reported from Chile, an estimated 7,256 tU could be recovered as a byproduct resource in the active Chuquicamata copper deposit and in the Bahia Inglesa and Mejillones phosphate deposits (NEA–IAEA, 2006).

Gabon produced uranium before 2006; the 2009 "Red Book" reports 4,800 tU in the second highest cost category. Areva and Pitchstone Exploration Ltd. have active exploration campaigns in Gabon (NEA–IAEA, 2010; Ux Consulting Company LLC, 2010).

Indonesia has never produced, nor is it currently producing, uranium, but it reports 4,800 tU at <USD 130 kg/U (NEA–IAEA, 2010). Although uranium deposits in Indonesia are well known, exploration is minimal and development is not likely in the foreseeable future (Dahlkamp, 2010).

The 2009 "Red Book" credits Italy with 4,800 tU in the <USD 130 kg/U cost category (NEA–IAEA, 2010). Italy passed a series of legislative packages during 2008 through 2010 that emphasized the importance of nuclear power. An assessment of prospective ground in the northern Alps near Switzerland was undertaken in an attempt to stimulate foreign investment (Vettraino, 2010). In 2011, following Fukushima, the nuclear policy of Italy was effectively reversed, ceasing investigations at least through 2016 (Vettraino, 2011).

The 2009 "Red Book" reported Japan as having 6,600 tU in the cost category of <USD 130 kg/U (NEA–IAEA, 2010). Japan has historically produced minimal uranium. Known resources are attributed to the Tono and the Ningyo–Toge deposits (Dahlkamp, 2010). Pilot-test mining of Ningyo–Toge during 1964–1982 yielded a small amount of uranium (87 tU), and this facility has been dismantled. Underground development of the Tono deposit was initiated, but this deposit is not active. Japan imports all its uranium, and Japanese utilities are involved in exploration and mining ventures worldwide. Japan has been on the forefront of developing technology for extracting uranium from seawater.

Peru reported 1,300 tU in the cost category of <USD 80/kgU in the 2009 "Red Book". This resource is from the Macusani area, currently being explored by Macusani Yellowcake Inc., Mega Uranium Ltd., South Andes Energy Inc., and Vena Resources Inc. (Vena) (Mega uranium Mining Company, 2011). Vena reports a 7,997 tU reserve for their deposit in the Macusani area, jointly held with Minergia S.A.C. This resource is still being explored, and has not progressed to the feasibility stage (Ux Consulting Company LLC, 2010). Peru has favorable geologic attributes for uranium mineralization, and exploration groups are active in some districts. The possibility of this country becoming a producer in the near future continues to increase as active exploration proceeds.

Portugal reported 4,500 tU in the cost category of <USD 80/kgU, and an additional 1,500 tU in the higher cost 2009 "Red Book" category of <USD 130/kgU(NEA–IAEA, 2010) . Portugal has produced uranium, but not since 2001, and nuclear energy is not considered in the current national energy policy. Some private companies have expressed interest in exploring the Nisa uranium property, but none has yet been granted mineral rights (NEA–IAEA, 2010).

The IAEA Secretariat estimated that Slovenia contains 1,700 tU in the 2009 "Red Book"'s cost category of <USD 130/kgU in 2009 (NEA–IAEA, 2010). This resource is attributed to the Zirovski deposit, owned by the Republic of Slovenia, which the government decided to decommission in 1994. Slovenia has not had recent or ongoing exploration or mine development activities, and is unlikely to contribute to the world uranium supply in the near future.

Sweden reported 400 tU in the cost category of <USD 130/kgU in 2009 (NEA–IAEA, 2010). Since 2007, exploration has increased in Sweden, with black shale deposits containing a very large potential resource. The Viken deposit is estimated to contain an indicated resource of 20 million tU, and inferred resources of 2.4 billion tU (Puritch and others, 2010). The metallurgy of this deposit precludes easy recovery, however, and Continental Precious Minerals Inc., the company currently exploring the Viken deposit, is reportedly investigating the use of bioleaching technology to leach uranium from these shales. Mawson Resources Ltd. is exploring three deposits of particular note in northern Sweden: the Hotagen unconformity related deposit (1,270 tU); the Tasjo uranium rare-earth deposit (42,300 tU); and the Dubblon volcanogenic deposit containing 3,366 tU (Mawson Resources Limited, 2008; Ux Consulting Company LLC, 2010). Aura Energy Ltd. is exploring low-grade black shales in its Haggan deposit in central Sweden with a reported resource of 114,642 tU, which the company feels might be profitable if coproducts molybdenum, vanadium, nickel and zinc are mined. Sweden is highly prospective, although the black shale and phosphate deposits will require either improved technology or the production of coproducts, or both, in order to be feasible (Ux Consulting Company LLC, 2010). No deposits are yet at the feasibility stage in Sweden, although the active exploration activities in the country may advance these deposits to feasibility in the near future.